This book is a short guide for the land surveyor and civil engineer to the world of the hydrographic surveyor. Activity offshore has been centred, until recent times, almost entirely on the fisheries and shipping industries. In the post-war years there has been a remarkable increase in interest in the resources of the sea and sea floor, of which the recovery of hydrocarbons – oil and gas – represents by far the greatest industrial investment. Offshore engineering technology has progressed at a phenomenal rate.

The author has written the book in the hope that the reader may be helped in his search for the scant definitive literature available and assisted in the attempt to rationalise the many facets of measurement and investigation in the marine environment.

Lt Commander A E Ingham RN(Retd) is Senior Lecturer in Hydrographic Surveying at North East London Polytechnic

Hydrography for the Surveyor and Engineer

Aspects of modern land surveying

Series editor: J. R. SMITH, A.R.I.C.S.

OPTICAL DISTANCE MEASUREMENT
by J. R. Smith, A.R.I.C.S.

MODERN THEODOLITES AND LEVELS
by M. A. R. Cooper, B.Sc., A.R.I.C.S.

ELECTROMAGNETIC DISTANCE MEASUREMENT
by C. D. Burnside, M.B.E., B.Sc., A.R.I.C.S.

DESK CALCULATORS
by J. R. Smith, A.R.I.C.S.

FUNDAMENTALS OF SURVEY MEASUREMENT
AND ANALYSIS
by M. A. R. Cooper, B.Sc., A.R.I.C.S.

HYDROGRAPHY FOR THE SURVEYOR
AND ENGINEER
by Lt. Commander A. E. Ingham, R.N.(Retd), A.R.I.C.S.

Hydrography for the Surveyor and Engineer

LIEUTENANT-COMMANDER A. E. INGHAM R.N.(Retd), A.R.I.C.S.

Senior Lecturer in Hydrographic Surveying
Department of Surveying
North East London Polytechnic

A Halsted Press Book

JOHN WILEY & SONS
New York

Published in the U.S.A. by Halsted Press,
a Division of John Wiley & Sons, Inc. New York

Library of Congress Cataloging in Publication Data

Ingham, Alan E
Hydrography for the surveyor and engineer
(Aspects of modern land surveying)

"A Halsted Press book."
Bibliography: p. 124
1.   Hydrographic surveying. I. Title.
VK591.147   526.9'9   73–21791

ISBN 0 470–42 730–2

Printed in Great Britain

Preface

This book is a short guide for the land surveyor and civil engineer to the world of the hydrographic surveyor. No claims are made, either for the completeness of the treatment or the originality of the text. Rather, it is hoped that the reader may be helped in his search for the scant definitive literature available and assisted in the attempt to rationalise the many facets of measurement and investigation in the marine environment.

The co-operation received in the compilation of my textbook *Sea Surveying* (John Wiley, 1974) has naturally benefited the writing of this work also. Full credit for this invaluable assistance would need a separate volume which might serve as a directory of those involved in sea surveying in all its aspects. Special mention must be made, however, of those most closely involved in the writing of this book, and particularly the staff of Decca Survey Limited, Sercel and Tellurometer (UK) Limited; Dr J. C. Cook, whose advice on the acoustics chapter was invaluable; Mr R. Ekblom, who checked so thoroughly the entire script; and Mrs R. Warren, who typed the final draft and helped to translate into readable English the meanderings of my ungrammatical mind.

To all of these persons and organisations I offer my sincere gratitude while, at the same time, accepting full responsibility for errors and omissions.

North East London Polytechnic   Alan Ingham
January, 1974

Contents

Introduction

Activity offshore has been centred, until recent times, almost entirely on the fisheries and shipping industries. Since World War II there has been a remarkable increase in interest in the resources of the sea and seafloor, of which the recovery of hydrocarbons—oil and gas— represents by far the greatest industrial investment.

Offshore engineering technology and the shipping industry have progressed at a phenomenal rate. Compare the typical, long-established harbours with the civil engineering achievements of today. The one occupies a sheltered location adjoining shallow waters, comfortably able to accommodate vessels which seldom exceed 10 m in draught, and conveniently close to road and rail distribution links. The other may be a drilling rig or production platform 150 km offshore and subjected to violent storms and 20 m high waves, or an exposed tanker terminal with depths exceeding 25 m alongside, or a pipeline from oil well to shore, traversing areas of rapid current and shifting seafloor sediments.

In addition to offshore drilling and harbour construction, the totality of industrial activity embraces the following operations:

dredging, for harbour conservancy, mineral recovery and reclamation;
coast protection engineering;
salvage;
desalination of seawater to improve fresh water supplies;

the extraction of minerals and chemicals from seawater;
the provision of recreational facilities such as beaches and marinas;
the prevention or elimination of pollution;
the development of communications and distribution routes by
    shipping lane, submarine cable and underwater pipeline;
the development of the fishing industry.

The impact of this proliferation of activity on the engineer and surveyor has been profound.

Seawater is a notoriously corrosive substance and this, together with the forces of current, tide and storm sea and the great pressures on structures at depth, constitutes formidable factors with which the engineer must contend in the marine environment. For him to do so successfully these factors must be quantified.

As on land, environmental data are acquired by the geologist and geophysicist, who base their work on the surveyor's map. At sea, in addition, the oceanographer plays his part in determining the sea-water parameters. All of these turn to the hydrographer for information on the physical limits of the marine environment. The data supplied by the hydrographer, and his function in the sea surveying operation, are illustrated in Fig. 1. It should be noted that there is a considerable overlap of scope and function of the specialists concerned, each being essentially a *team*-member working with the others towards a common goal. Hydrography is similar in many respects to land surveying and many of the techniques employed are the same as, or extensions of, land surveying practice. Some factors serve to differentiate between the two, however, and whilst some are obvious, all must be appreciated and taken into account.

The chart is the marine equivalent of the topographic map. Both use spot measurements of height/depth and contours to portray relief, but where the user of the map is able to verify by visual inspection the detail shown, the seafloor topography is obscured. The chart-user therefore relies implicitly on the accuracy and thoroughness of the hydrographer's work.

In establishing the control for a topographical survey the land surveyor occupies a number of stations in turn, observing angles and measuring distances with all the care, precision and number of repetitions required to achieve the specified order of accuracy. The hydrographic surveyor likewise performs this control procedure in most instances. Once afloat, however, the fixation of position becomes a dynamic operation. Not only does the observing platform (the survey

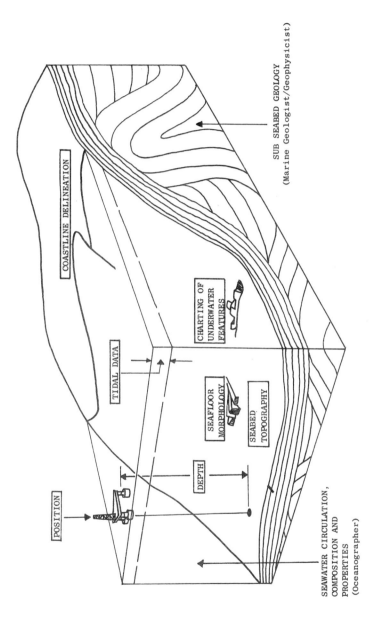

Fig. 1. The data requirements of offshore industry (Those with which the hydrographer is commonly concerned are shown framed)

vessel) occupy a position for an instant only, but the level of the sea-water surface changes constantly under the influence of tide and wave. The hydrographic operation has been likened to a levelling survey onshore, but using a telescoping staff and a level which is mounted on a well-sprung trolley! Admittedly this type of surveying is less precise than the shore-based equivalent, but this does not mean that less careful, less thorough work is adequate. On the contrary, the utmost care is essential to ensure that the rather more coarse measurements which are combined to make the chart are not further degraded by slapdash methods.

The nature of the sea environment is probably the most fundamental single factor which separates land from sea surveying. The effect of the sea on the common surveying techniques has been mentioned, but more important still is an appreciation of the vicissitudes of the sea which not even a manual of seamanship can properly explain. Experience is the only real solution, which both engineer and land surveyor will most probably lack. In its stead, an honest humility towards the sea and the seaman in whose care the surveyor will be placed is strongly recommended.

Always bearing this sobering thought in mind, it is the intention in this book to give the reader having a familiarity with surveying ashore an insight into the special techniques and problems of surveying afloat.

CHAPTER ONE

The Elements of Hydrography

The first object of a hydrographic survey is to depict the relief of the seabed, including all features, natural and manmade, and to indicate the nature of the seabed in a manner similar to the topographic map of land areas.

Two factors define the location of a single point on the earth's surface taken in isolation, and for the sea survey these are:

(i) the position of the point in the horizontal plane in, for example, latitude and longitude, grid co-ordinates or angles and distances from known control points;

(ii) the depth of the point below the sea surface, corrected for the vertical distance between the point of measurement and water level and for the height of the tide above the datum or reference level to which depths are to be related.

The problem, then, is how to apply these factors in order to obtain a pictorial representation of the seabed relief.

In surveys ashore, photogrammetric techniques apart, contours are derived from an accumulation of spot heights obtained by levelling or other means, often in a grid system. The lead-line survey would be exactly analogous, except for the necessary allowance for tidal height which makes each depth measurement unique in time. A depth thus measured is termed a *sounding*. When the corrections have been applied, the depth is a *reduced* sounding. (See Fig. 2.)

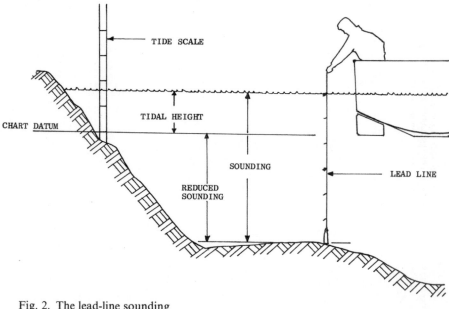

Fig. 2. The lead-line sounding

In fact, the lead-line is rarely used and the process is streamlined by the echo-sounder. The echo-sounder effects a depth measurement typically at a rate of up to 6 times per second, depending on the depth. The measurement is made by timing the interval between transmission and reception of an acoustic pulse which travels from the vessel to the seabed and back at a velocity of approximately 1500 m.s^{-1}. A succeeding transmission is not made until the previous pulse has returned and the rate of transmission is thus depth-dependent. The vessel is therefore able to proceed without stopping and, by sounding continuously, a profile is obtained of the seabed beneath the vessel's track. We then speak of a *line of soundings* having been run. In practice, a succession of parallel sounding lines is run across the survey area and contours are derived from the resulting profiles to build up a portrayal of the seabed topography.

In the most highly automated survey systems, position fixes are obtained at a rate compatible with the sounding rate, but this is not, as yet, a common situation. Instead, a fix may be obtained at intervals of a minute or so, and the survey vessel is assumed to follow a track which is shown on the plot as a succession of fixes joined by straight lines. It is imperative that control of the vessel is such that this

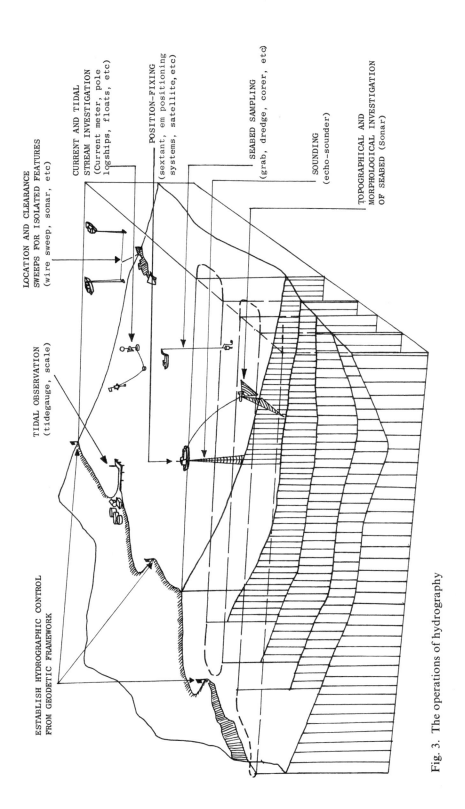

LOCATION AND CLEARANCE
SWEEPS FOR ISOLATED FEATURES
(wire sweep, sonar, etc)

CURRENT AND TIDAL
STREAM INVESTIGATION
(Current meter, pole
logships, floats, etc)

POSITION-FIXING
(sextant, em positioning
systems, satellite, etc)

SEABED SAMPLING
(grab, dredge, corer, etc)

SOUNDING
(echo-sounder)

TOPOGRAPHICAL AND
MORPHOLOGICAL INVESTIGATION
OF SEABED (Sonar)

TIDAL OBSERVATION
(tidegauge, scale)

ESTABLISH HYDROGRAPHIC CONTROL
FROM GEODETIC FRAMEWORK

Fig. 3. The operations of hydrography

assumption is valid within the plottable accuracy of the chosen scale of the survey. (See Chapter 3, para. 3.4.)

The sounding procedure described above, while providing ample data for an indication of relief, cannot guarantee a complete coverage of the seabed. Isolated pinnacles, wrecks and other obstructions may be missed if they lie between the sounding lines. Further, the echo-sounder profile will not show the nature of the seabed—where rocks outcrop from sand, where gravel, stones or boulders occur, and so on. The conventional sounding operation may therefore be modified—by running intermediate lines (interlines) to increase the density of areal coverage or cross-lines to obtain an improved angle of cut of contours—or supplemented by special instrumentation, such as seabed samplers and the side-scan sonar, or techniques such as wire sweeping. Additionally, tidal and tidal-stream observations are frequently specified. These various operations are dealt with in succeeding chapters, while their roles in the hydrographic survey are illustrated in Fig. 3.

CHAPTER TWO

Planning

2.1 General

As with any activity, careful planning and preparation will pay dividends in the subsequent operation. Although no two surveys are alike, the sequence of events will usually follow much the same pattern:

(i) the drawing up of a specification to the client's requirements;

(ii) the examination of available documents, e.g. charts, maps, air and ground photography, sailing directions, tide tables, triangulation and other control data from earlier surveys;

(iii) if at all possible, a field reconnaissance;

(iv) the preparation of sheets required for the survey, e.g. master plotting sheet, sounding boards, underlays for position-fixing lattices, and the fair sheet;

(v) the preparation of the operational plan, including decisions on positional control, tide gauge location, instrumentation and techniques to be used, personnel, equipment and logistical requirements and time/resources schedules;

(vi) the field work, e.g. establishing positioning and tidal control, sounding, sweeping and miscellaneous operations;

(vii) the interpretation, processing and presentation of data.

The above list will be quite familiar to the land surveyor, only the manner in which each phase is executed being peculiar to the hydrographic operation.

It must be self-evident that the client, probably unacquainted with such operations afloat, will need to be educated accordingly (albeit tactfully!) so that he does not demand the impossible and pay dearly for it. Every item in the list must be approached with a marine-orientated eye, and possibly the example most vividly illustrative of this is that of cost.

Every aspect of hydrography is more costly than its land survey equivalent. An observing team ashore, for example, might consist of four men travelling by Land Rover and using two theodolites. For a sounding survey a team of perhaps only three men will be required, but, according to size, the boat may cost between twice and one hundred times more than the Land Rover, with correspondingly higher depreciation and running costs and a crew of between one and twenty men. An echo-sounder is essential at a cost of at least two theodolites, while the minimum of two sextants for position-fixing are equivalent to yet another. The Land Rover will travel at, say, 30 m.p.h. to observing points which can be quite conveniently located relative to the base camp. The vessel's speed will most likely be around 10 knots and the survey area is frequently as far as 100 miles from the nearest harbour. The logistics—victuals, fuel, passage time between base and survey area—will probably play a much larger part in the scheduling of the survey afloat, and the effects of weather conditions on the progress of the survey will be more marked. The 'domestic' arrangements for a hydrographic survey are therefore as important as is the achievement of the required standards of accuracy and, whilst every care should be taken, it is always wise to aim for results which are *adequate* rather than those which are the best attainable.

Flexibility of plan is a desirable attribute, so that the effects of delays due to breakdowns, weather, sickness and the numerous other possible causes can be assessed and the correct remedial action taken at once. For this, the 'critical path' or network plan is highly recommended. Finally, though the survey at sea is more costly than similar work on land, the surveying phase is less costly by far than any succeeding phase of offshore activity, thus placing great reliance on the integrity of the surveyor.

The factors to be considered in the sounding operation are as follows:

(i)   type of vessel required;
(ii) type of echo-sounder/sonar instrumentation required;
(iii) position-fixing method to be used;

(iv) method of achieving the required coverage of the seabed;
(v) personnel requirements;
(vi) logistical requirements.

2.2 The Survey Vessel

While many surveyors will be committed to a particular vessel, there will be those who must charter a vessel in the locality of the survey area. The most suitable craft for the job will obviously depend on the overall purpose of the survey (e.g. whether or not geophysical or other additional tasks are required), the weather conditions to be allowed for, the size of the survey team, whether they are to live on board the vessel, and so on. It is impossible to lay down rules, but the following requirements will always apply:

(i) The vessel should be spacious enough to allow for plotting and fixing, the former preferably under cover and free from engine vibration.
(ii) If fixing by visual methods, an all-round view is required by the anglers.
(iii) The vessel should be stable and manoeuvrable at slow speeds.
(iv) Unless batteries or a generator are to be provided by the survey team, there should be ample electrical power for all foreseeable needs.
(v) The range (fuel capacity), food storage and working facilities should be compatible with the planned operational arrangements.
(vi) Speed capability need not normally exceed 10 knots or so, though the distance between the base and the survey area may dictate a faster vessel to avoid undue loss of time on passage.

2.3 The Sounding Plan

In order to achieve satisfactory coverage of the seabed with maximum economy it has already been explained that parallel lines of sounding should be run over the area (see p. 6). Further considerations will include the following, all of which are interrelated:

(i) the scale of survey appropriate to the precision and thoroughness required;
(ii) the spacing apart of the sounding lines;
(iii) the interval between fixes along a line;

(iv) the speed of the vessel while sounding;
(v)   the direction in which lines are to be run.

Ship speed will rarely be limited by the data acquisition rate of modern echo-sounders, though in very high-speed surveys it may be necessary to compensate for the Doppler frequency shift between transmitted and received pulses. The position-fixing method is much more likely to dictate the speed at which the sounding lines are run, and the following example illustrates the interaction of the factors involved.

Let us suppose that a survey is required of a harbour on a scale of 1:10 000. Position-fixing is to be by sextant angle resection, plotted by station-pointer. A fixing interval of one minute is reasonable for these circumstances. Now, the vessel will be assumed to follow the straight line joining the fixes taken every minute, and it follows that the fixes should be spaced not too far apart for such an assumption to be valid. The Royal Navy Surveying Service relates the scale of the survey to two rules:

(i) fixes should never be further than 25 mm apart on the scale of the survey, and
(ii) the spacing apart of sounding lines should not exceed 10 mm.

Thus, by setting the scale of the survey at a particular value, a certain thoroughness (density of soundings or areal coverage) and precision is implied. If the above rules are followed in our example, fixes will lie 250 m apart. This, together with a time interval of one minute, decrees a vessel speed of 250 m.min^{-1}, or 15 km.h^{-1} (about 8 knots). (A scale of 1:5000 implies a sounding speed of 4 knots, and so on.) To obey the rule at a higher sounding speed, a more rapid method of position-fixing is required.

The sounding lines will normally be run in a direction as nearly as possible at right-angles to the expected direction of the depth contours at a spacing of 100 m. Here again, we are able to assess the degree of coverage given by the echo-sounder transmission. The acoustic power is focussed into a beam which will emanate from the vessel in a cone shape. A typical beam width is about 30° (see Chapter 4) and we can therefore calculate that the 'insonified' area of seabed beneath the vessel will meet that of the adjacent lines only in depths exceeding 185 m (Fig. 4). It is then left to the surveyor to decide whether the gap between lines in depths less than this can be accepted, or whether to increase the scale, decrease the sounding line spacing, or use additional

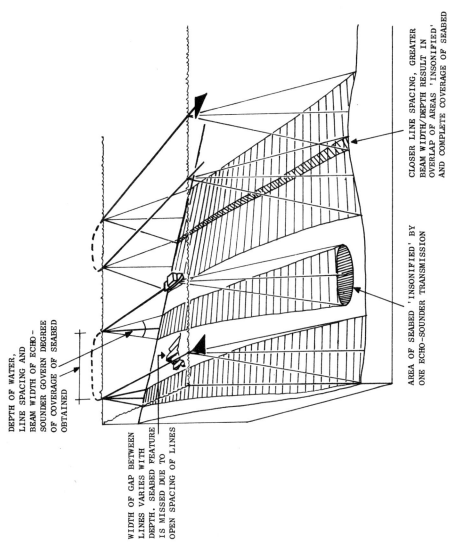

DEPTH OF WATER,
LINE SPACING AND
BEAM WIDTH OF ECHO-
SOUNDER GOVERN DEGREE
OF COVERAGE OF SEABED
OBTAINED

WIDTH OF GAP BETWEEN
LINES VARIES WITH
DEPTH. SEABED FEATURE
IS MISSED DUE TO
OPEN SPACING OF LINES

AREA OF SEABED 'INSONIFIED' BY
ONE ECHO-SOUNDER TRANSMISSION

CLOSER LINE SPACING, GREATER
BEAM WIDTH/DEPTH RESULT IN
OVERLAP OF AREAS 'INSONIFIED'
AND COMPLETE COVERAGE OF SEABED

Fig. 4. The interrelation of survey scale and sounding line spacing with the instrumentation employed

instrumentation such as the side-scan sonar to investigate the seabed between lines for features which may require attention.

In the example above, the survey scale has been defined, and all other factors have been calculated from it. The converse may be the case. For example, the echo-sounder beam width of 30° is required to give complete seabed coverage in general depths of about 30 m. In this case, line spacing should be 15 m and, for this to be represented by about 10 mm on the scale of the survey, the scale should be 1 : 1500. The spacing of fixes at 25 mm on paper is therefore 37·5 m. If a fix is possible every minute, the sounding speed will be 2·25 km.h^{-1} or about 1·2 knots. It is quite possible that at this speed the vessel will not be controllable (i.e. it will have lost 'steerage-way') and a more rapid fixing rate will be required.

2.4   Tidal Control

Tides and tidal streams are a most important part of the hydrographer's environment and a full study should be made of the subject by reference to such volumes as the *Admiralty Manual of Tides* and the *Admiralty Manual of Hydrographic Surveying*, Vol. II. In particular, the surveyor is frequently required to obtain tidal and tidal stream observations as raw data for analysis to determine the constituents of the astronomical tide-generating force. These constituents then form the basis of predictions of the heights, ranges, rates and directions of the future. Admiralty Tidal Handbooks Nos. 1, 2 and 3 are the standard works for these matters.

For the sounding operation, tidal information is necessary only for the reduction of soundings to a datum. Again, tidal datums are fully explained in Admiralty Tidal Handbook No. 2, *Datums for Hydrographic Surveys*. The requirement is twofold. On the one hand, a suitable datum must be determined, and on the other, observations of tidal height above the datum are required throughout the sounding operation for the reduction of soundings.

The sounding datum may be a totally arbitrary level, always expressed as a certain distance below a permanent mark such as a benchmark of the land levelling system (e.g., in the UK, a benchmark whose height is known in relation to Ordnance Datum, Newlyn). An engineer might, for instance, choose a datum at the level of the cill of a proposed new dock, but it is usual practice to relate the datum to the tide at a place. Thus, sounding datum is usually fixed at the level of

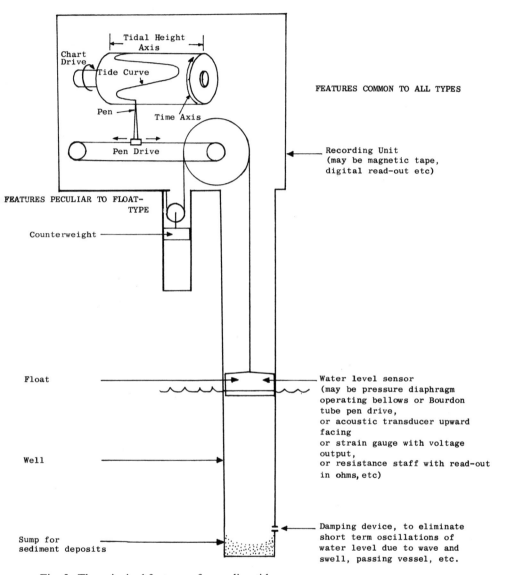

Fig. 5. The principal features of recording tide gauges

the lowest predicted tide, and since predictions are based chiefly on the astronomical tide-generating forces, this is the level of Lowest Astronomical Tide (LAT). In most places in developed countries it is possible to recover LAT from existing data and established benchmarks. It may sometimes be necessary, however, to establish a new benchmark and observe the range of the tide over a period to determine a suitable datum level. Full instructions are contained in the Handbook.

Once the datum level has been selected, a tide pole or gauge (known collectively as 'tide scale') is required for tidal information during sounding operations. A simple graduated pole, erected with its zero mark below datum level and in length sufficient to accommodate the range of the tide, is perfectly adequate, but an observer is required to record the heights of tide and this may prove uneconomical. The recording tide gauge spares the observer's services and provides a permanent record, and its cost can usually be justified on these grounds. The various types are illustrated in Fig. 5.

The most frustrating aspect of tidal observation is the difficulty of locating the observation point close enough to the survey area. The tidal range, indeed the whole regime, can vary quite radically from place to place and geographical factors (e.g. slope of seabed, configuration of coastline, impounding of water due to bars, islands, etc.) can produce significant differences between the tide as recorded at a gauge in a sheltered harbour and as experienced in an area close offshore. Much effort has been expended over the past few years to produce a satisfactory seabed gauge, with little success, though a few are available which give fair results.

However, the problems still remain of ensuring the stability of the gauge mounting on the seabed and of relating the gauge zero to a permanent reference mark. In the absence of a seabed gauge, co-tidal charts must be used for computing the reduction in areas offshore, and these are far from satisfactory in view of the paucity of tidal information on which they are based (see Admiralty Tidal Handbook No. 2).

CHAPTER THREE

Position-fixing Afloat

3.1 General

A major element of the hydrographic survey is the dynamic positional control of the survey vessel, and until the method to be used has been chosen, the sounding speed, time schedules, costs, etc., cannot be assessed.

As in land surveying, the position of a point is determined relative to other, known points by the intersection of a minimum of two position lines. These may be lines of bearing, arcs of range, angles constituting an intersection or resection fix, or combinations of any of these. The reference points (*shore stations*) on which offshore fixes are based are fundamental to any positioning method and are usually the hydrographer's first concern.

Land surveying methods are used to establish the control required, e.g. geographical (astro) position fix, azimuth determination, base measurement, triangulation, and so on. More usually, an established control network will be used, position being transferred to the new stations required by intersection, resection, traverse, etc. The number and siting of these stations, and the precision with which they will be co-ordinated, depends largely on the method selected for the offshore fix.

In general, and with the notable exception of the satellite fix, the longer the range, the lower the resolution of the dynamic methods and the lower the standard of accuracy achieved. Moreover, the longer the potential range of a method, the fewer are the shore stations required and the greater the precision with which they must be fixed. In the case of the long range and world-wide systems the shore control is out of the hands of the surveyor altogether, whereas his involvement in the control for short and medium range methods is total.

For all short range methods, the shore control stations should be as near as possible to the shoreline, so that the measurement of slope-angles or ranges from elevated stations is averted.

Track control is an important factor in position-fixing afloat, ensuring that the sounding line pattern is neat and gives methodical coverage of the survey area, and that the vessel does not depart unduly from the intended lines.

3.1.1   Methods of Position-fixing and Track Control

Optical Methods
(i)   Double horizontal sextant angles observed simultaneously to obtain a resection fix. This is the most common and versatile method in overall use, particularly at distances of from 200 m to 5 km from shore marks. It is relatively inexpensive, a sextant costing in the region of £200, and station-pointers for plotting about £50. Accuracy will usually be in the region of ± 3–5 m within the above range bracket.
(ii)   Single vertical sextant angle subtense (for conversion of angle to range) combined with a second cut-off angle or range. For use in large scale surveys out to about 200 m from the shore marks. A subtense board is required. This may be a pole of fixed length, the observed vertical angle being converted to range away from the pole, or, better, a pole marked along its length at intervals representing the base line subtended by a fixed vertical angle at the required intervals of range (Fig. 6).
(iii)   Intersection from theodolite stations ashore. This is not to be recommended, since complex arrangements are required to ensure that both angles are measured at the instant of the fix marked on the echo-sounder record, and to communicate the fix data to the boat. Plotting of such a fix is also a cumbersome operation.
(iv)   Compass and range finder. These and similar observations offer only low standards of accuracy and plotting is cumbersome. As the sextant is an essential tool of the surveyor it is difficult to justify the use of such imprecise alternatives.

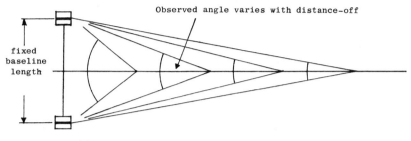

(a) Subtense board of fixed length

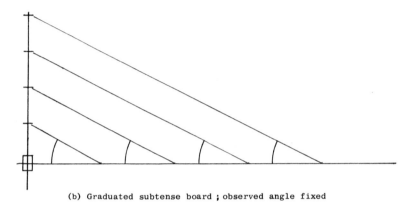

(b) Graduated subtense board ; observed angle fixed

Fig. 6. The subtense methods of range measurement

Direct Method: Distance Line
(v)   This is simply, as the name implies, a line marked along its length which is stretched between vessel and shore whereby the distance-off may be measured as required. As these distances are necessarily short, the method is usually only applicable to soundings along a wharf or pier, and the subtense methods are almost always preferred.

Radio Methods
(vi)   'Line-of-sight' microwave electromagnetic position-fixing (EPF) systems. Using carrier frequencies in the order of 3000–10 000 MHz, they are very similar to the microwave EDM instruments used in land surveying, such as the Tellurometer. Accuracies of ± 2 m are normal, fixing by two simultaneously measured ranges, at up to 40 km or so from the remote instruments ashore. Costs are naturally very much

higher than for visual methods, £15 000–20 000 being typical of the price for a two-range system complete.

(vii) Medium range EPF systems. Operating in the 1·5–4 MHz frequency band with range capability of 150–1200 km depending on power output, these systems employ either the two-range or hyperbolic modes of operation. They are transportable and widely used.

(viii) Long range EPF systems. From 150 km range to virtually world-wide operation, these low—and very low—frequency systems are unlike the shorter range versions in that the transmitters occupy permanent locations and operate only in the hyperbolic mode.

(ix) Satellite navigation (SATNAV) systems. A number of navigational satellites in polar orbits are used extensively ashore and afloat for precise position-fixation. The TRANSIT method employing Doppler techniques is internationally favoured for position-fixing at sea and its application is, of course, world-wide.

Miscellaneous Methods

Among these might be included radar, underwater acoustic position-fixing and laser ranging and direction indication.

(x) Radar is an old and well-tried system for target location with reasonably accurate ranging and only fair bearing discrimination. With certain specialised ranging attachments and transponder beacons placed at co-ordinated marks onshore, two-range fixing is possible to surveying standards at ranges up to 40 km.

(xi) Underwater acoustic beacons are much used for specialised purposes such as oil-rig well-head markers, dynamic ship-positioning systems over a fixed point on the seabed, and as seabed markers for checking the lane-count (and therefore the stability) of medium and long range electromagnetic systems. Various acoustic positioning systems with sonar ranging are available, whereby networks of seabed transponder beacons are placed, and then used, as a triangulation scheme might be used on shore, for intermediate positioning within the control net.

(xii) The laser has great development potential, both as a horizontal positioning aid and as a means of shallow depth measurement. Its most obvious use to the surveyor is as a directional beam along which a vessel might be steered, for track control. Ranging lasers are not yet widely used, and the movement of the vessel raises difficulties concerning the stability of the necessary reflectors in the highly focussed light-path.

The acoustic methods are described in para 4.4. For reports on new developments, which might include the laser and infra-red instrumentation, journals such as the *International Hydrographic Review* and *Offshore Services* are recommended.

The above list indicates a bewildering array of possible choices but the surveyor will rarely, if ever, be confronted with the entire range. An appreciation of the aim will usually make the selection quite unequivocal. For example, a small contract survey firm might rely on a couple of sextants, a portable echo-sounder and a launch hired locally. Work beyond 5 km offshore for such an organisation will be uncommon. The larger contracting company might operate its own vessels and use a hired medium range EPF system, while detached parties might carry out smaller surveys using a portable microwave system or sextants from locally hired craft. Geophysical organisations are much more likely to work far offshore with their own vessels, probably equipped for satellite positioning with Doppler sonar or inertial track-following instrumentation. Port survey departments may use a permanent medium range EPF system or portable microwave equipment, depending on size, but always with sextant fixing at least as a stand-by method.

3.2   Sextant Fixing

The sextant is not as familiar to the land surveyor as it deserves to be. For rapid observation with a resolution of ± 0.5 minutes of arc, a properly calibrated sextant is invaluable, either for vertical angles (in the surveying context, usually for subtense work) or horizontal, provided the observer may be confident of the measured angles being truly horizontal.

The sextant and its calibration is described in many works, notably the Admiralty manuals of hydrographic surveying and navigation. It will serve here to mention that the index and side errors should be frequently checked and corrected, while parallax and collimation errors should be guarded against.

The resection fix is no stranger to the land surveyor. In its dynamic role, however, it is doubly important to ensure that the fix is sound and easy to observe. The angle of cut of the position arcs of the fix should naturally be large—as near $60°$ as possible—and in sextant angling this implies angles of between $20°$ and $110°$ ideally, always summing to more than $50°$. The danger circle, when the vessel and all three shore marks lie on a common position circle, is a very easy trap for the unwary to fall into and the selection of marks for the fix should take particular account of this.

For precision, the two anglers should be placed as near as possible to each other since the fix will be plotted as though observed from a single point. Furthermore, the anglers should observe from directly above the point of measurement of depth.

The sextant resection fix is plotted either by station-pointer, or by eye using a plot of fixed subtended angles. The station-pointer is simply a 360° protractor with legs pivoting about the centre which can be set to the observed angles of the fix. By placing all three legs on the plotting sheet over the stations observed, the ship's position is found at the centre of the protractor. Plotting by station-pointer is rather cumbersome and practice is necessary to develop a reasonable plotting rate. An experienced surveyor should be able to observe one angle of a resection and plot the fix by station-pointer in about 30 s. A further 10–15 s should be allowed to prepare for the next fix, implying a normal interval between fixes for this method of one minute.

The calculation and plotting of arcs of constant subtended angle is simple (though time-consuming) and is fully described in the *Admiralty Manual of Hydrographic Surveying*. They may be justified in almost every case and their advantages are numerous. The station-pointer may be dispensed with, and this in a small boat may be a major benefit. The angle of cut of position for a chosen fix can be seen at a glance. The angles which constitute a fix need not be adjacent, thereby increasing the flexibility of the method. The angles of the next fix can be predicted easily and accurately and the moment when a fix ceases to be sound is immediately apparent. The disadvantage often quoted, that the selection of marks is limited because a proliferation of 'families' of arcs is confusing, can be easily overcome by preparing several alternative transparent underlays, each bearing only a few 'families'.

Probably the greatest advantage of all lies in the improved accuracy made possible by arcs of constant subtended angle. In sextant angling, the angle must be 'followed', i.e. the marks kept in coincidence by frequent movement of the micrometer screw. At the instant of the fix, it is almost impossible to guarantee that the movement is arrested in both sextants simultaneously, or that the angles have been observed within the minimum resolution of the sextants. If a constant angle plot is available, however, it is possible to arrange to fix at a pre-determined value for one of the angles. Thus a fix might be taken when one angle is, say, 30°, then 31°, 32° and so on. These values may be set on the sextant carefully in advance and no action is required at the instant of fix except to give the order. Then, if the vessel is steered around the other arc, so that the other angle is kept as near constant as possible, the minimum movement of the micrometer is sufficient to bring the appropriate marks into coincidence and complete the fix. Further, since the vessel is steering along that arc, the parallax effect due to the separation of the two anglers is minimised and the maximum possible accuracy is achieved. The elements of fixing by sextant angles are summarised in Fig. 7.

Subtense methods can simplify the sounding operation still further in certain circumstances. Parts of the channels and berths of many ports require regular dredging programmes and continual re-surveys of these areas are necessary. The information required concerns the depths and width of the channel for navigation or alongside and adjacent to the berths. Survey scales tend to be very large (e.g. 1:500 to 1:1500) and limited in extent.

The procedure adopted at one major port in these circumstances is as follows (see Fig. 8). The line-ends are permanently marked along the edges of the wharves and jetties. A subtense board is set up at one such mark. The board is marked at intervals which represent the ranges subtended by a fixed sextant angle of 5°, its zero positioned at the level of the height of eye of the observer in the survey launch. The direction of the line is defined as a sextant angle from a reference object, and the shore observer directs the launch along the line by flag signals or instructions by radio. The surveyor simply observes the subtense board from the launch through his sextant with 5° set on the arc, marking the echo-sounder record as each distance mark is brought into coincidence with the zero in turn. Thus, the launch is fixed by the intersection of the range-arcs and the sounding line direction.

It is appreciated that this method is based on the assumption that the launch does not stray from the pre-arranged sounding line and no check is kept of the accuracy of this assumption. However, the shore observer can instruct the launch crew to re-run an insufficiently precise line, and when the *trends* of changes in depth are more important than absolute accuracy the method is quite satisfactory.

The advantages are real. The parallax error caused by the separation of two anglers is eliminated. Furthermore, the fixes can be pre-plotted and may be sufficiently close together (at one second intervals if required) to provide one fix for each sounding plotted. Precision *along* the sounding line offered by the subtense ranges is high, and encroachment of silt into the channel, silting or scouring alongside

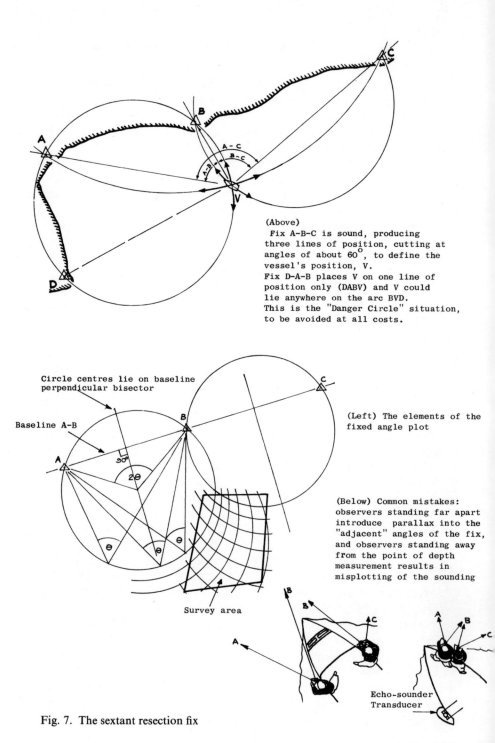

(Above)

Fix A-B-C is sound, producing
three lines of position, cutting at
angles of about 60°, to define the
vessel's position, V.

Fix D-A-B places V on one line of
position only (DABV) and V could
lie anywhere on the arc BVD.

This is the "Danger Circle" situation,
to be avoided at all costs.

Circle centres lie on baseline
perpendicular bisector

Baseline A-B

(Left) The elements of the
fixed angle plot

(Below) Common mistakes:
observers standing far apart
introduce parallax into the
"adjacent" angles of the fix,
and observers standing away
from the point of depth
measurement results in
misplotting of the sounding

Survey area

Echo-sounder
Transducer

Fig. 7. The sextant resection fix

wharves and the quality of a dredging operation are all revealed. Care must be taken to allow for the instrumental characteristic of the sextant wherein the index glass and telescope line of sight are separated by about 50 mm. This may be added to the calculated intervals of the markings of the subtense board. (See Fig. 8.)

3.3 Electromagnetic Position-fixing (EPF)

3.3.1 The theory of electromagnetic wave propagation and the principles of the distance measurement instrumentation used by surveyors are described in another book in this series, *Electromagnetic Distance Measurement* (Burnside: Crosby Lockwood, 1971). It is assumed here that the reader is familiar with, or has access to, that work.

As already stated, position-fixing afloat is a dynamic operation and in discussing the accuracy of EPF systems afloat it should be re-membered that the mobile—that is, the survey vessel or EPF system platform—is subject to the motions of pitch, roll, yaw and heave. The ship-borne antenna, usually at the masthead, is never still and can frequently sway through a radius of several metres from the point of depth measurement. A theoretical resolution of less than one metre is therefore somewhat pointless.

The vessel's position is defined by the intersection of two (or more) range circles centred on co-ordinated points ashore,* or of two hyper-bolae of range difference, and the issue is again complicated by the dynamic situation since the two ranges, or range differences, must be obtained simultaneously to give a fix at an instant.

The classification of EPF systems as short, medium or long range conveniently accords with the different characteristics of propagation of the frequencies employed. The equipment increases in size with wavelength, power requirements and thus with range capability. The short range systems are most similar to the land surveyor's EDM instruments—the Tellurometer for use afloat is indeed a dynamic version of, and little different from, its shore-based counterpart—and can be installed, transferred and dismantled at will in a very short time. The medium range systems are less readily portable but can still be considered as flexible in relation to their longer ranges, greater

---

* The 'shore' stations are sometimes placed on buoyed platforms offshore. Some accuracy is sacrificed due to the scope of the buoy mooring.

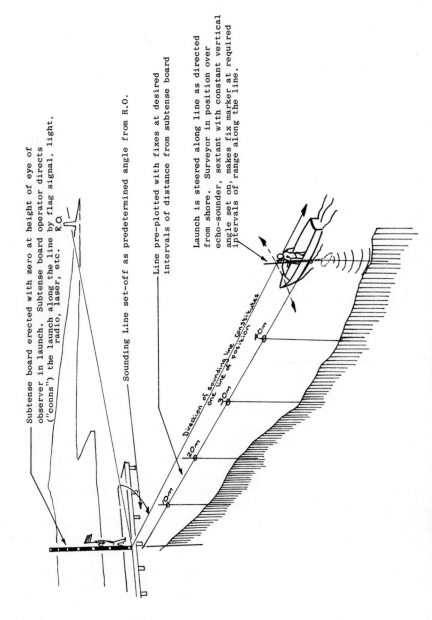

Subtense board erected with zero at height of eye of
observer in launch. Subtense board operator directs
("conns") the launch along the line by flag signal, light,
radio, laser, etc. R.O.

Sounding Line set-off as predetermined angle from R.O.

Line pre-plotted with fixes at desired
intervals of distance from subtense board

Launch is steered along line as directed
from shore. Surveyor in position over
echo-sounder, sextant with constant vertical
angle set on, makes fix marker at required
intervals of range along the line.

Direction of sounding line constitutes
one line of position

0m   10m   20m   30m   40m

Fig. 8. The subtense method of positioning for large scale surveys

areal coverage and the consequently longer useful 'life' of a particular installation.

The short and medium range groups can truly be considered as surveyors' tools, their initial placement, calibration and operation being under his control. The long range systems are permanently established and operated by national or international agencies, the user's equipment being passive (i.e. a receiver).

The circular lattices of the short range systems and the two-range modes of the medium range group are usually computed and drawn by the surveyor, allowing for projection scale factor where necessary. Although it is possible for hyperbolic lattices to be similarly hand drawn, the procedure is tedious and much more liable to error through the use of curves and splines. System manufacturers invariably supply users with comprehensive manuals and offer courses of instruction to user personnel and the services of trained engineers and surveyors for the setting up, calibration, operation and maintenance of a chain. Hyperbolic lattices, plotted automatically under computer control, can be included in this service.

While the read-out is basically visual (as an analogue or digital display of lane readings or distance), all modern systems are capable of binary coded outputs for computer processing and automatic print-out or plotting of fixes at required intervals. The most sophisticated incorporate the EPF components in a comprehensively automated system capable of steering the vessel along a required track, plotting the track continuously and even plotting soundings or geophysical data in real time. (See Chapter 6.)

In discussing EPF systems it is convenient to assume an approximate value for the velocity of propagation of e.m. waves (v) of 3×10^5 km.s^{-1}. Given the frequency of operation (f) it is then a simple matter to compute the wavelength $(\lambda = v/f)$ and thence obtain an indication of the resolution to be expected.

3.3.2 Short Range Systems

Short range equipment falls into the microwave frequency brackets, where range is limited to 'radio line of sight', or slightly in excess of visual line of sight, and may be from 25 to over 100 km depending on antenna height and output power. A velocity of propagation is assumed (based usually on an atmospheric refractive index of 1·000 33) and measuring frequency crystals are incorporated as appropriate to give a read-out direct in units of distance. Since the range information is required by the surveyor in charge afloat, the controlling, processing and read-out equipment is ship-borne. The vessel becomes the 'master' station while the 'remote' onshore may or may not be manned.

The actual measurement is performed indirectly, either by measurement of the phase difference between transmitted and returning c.w. signals (as in the Tellurometer), or by timing the travel of pulsed signals from master to remote and back (as in the Trisponder). The Tellurometer MRB 201 and the Trisponder 202A are two systems which typify these methods of measurement. The following notes are based on the manufacturers' literature.

The Tellurometer MRB 201 (Plate 1)
Using the well-known phase measurement techniques of the land surveying Tellurometers, the MRB 201 is the successor to the Hydrodist (MRB 2), used by hydrographers since 1960. One range is obtained from a master-remote pair, and the equipment must be duplicated for two-range operation. The directional paraboloidal reflectors with dipole antennae of the master and remote instruments must be orientated towards each other. The directional antennae may be separated from the instrument and placed at the masthead, or a duplex arrangement can be used to house both master antennae at the masthead with a facility for their mechanical orientation towards their remote instruments. A self-seeking antenna currently under development will automatically search for, lock onto and follow the remote station.

The master and remote instruments are identical and interchangeable. A digital range integrator (DRI) unit in the master instrument provides an integrated fine pattern reading which is updated every millisecond. In both the master and remote instruments a dial read-out unit can be fitted which allows static measurements to be made to a greater degree of accuracy. A memory is built into the MRB 201 so that, should the signal be lost temporarily, the slant range velocity is stored and recalled as necessary to provide a continuous read-out. An indicator is provided to inform the operator when true signals are not being received. The fine reading (A pattern) wavelength is 100 m (assuming an average refractive index of 1·000 33) and resolution in range to 0·1 m is possible within this 100 m bracket. Ambiguities within 1000, 10 000 and 100 000 m are resolved by combining readings from patterns D, C and B respectively in the manner of all Tellurometers. This is computed within the DRI circuitry by the

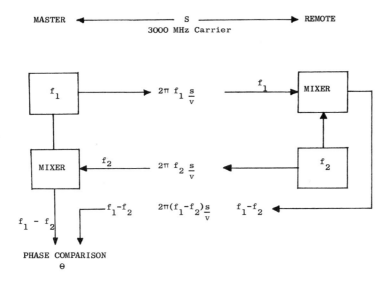

Fig. 9. The MRB 201 measuring principle

operation of a null-meter, resulting in unambiguous read-out of range to 100 km.

The measuring principle is as follows (see Fig. 9): a modulation frequency f_1 is transmitted on a 3000 MHz carrier from the master, and f_2 from the remote. The master and remote are separated by a distance s. If the instantaneous phases of f_1, f_2 are given by ϕ_1 and ϕ_2 respectively, then it can be seen that the final phase comparison θ will be given by:

$$\theta = \left[\phi_1 + 2\pi f_1 \frac{s}{v} - \phi_2 + 2\pi(f_1 - f_2)\frac{s}{v} \right] - \left[\phi_1 - \left(\phi_2 + 2\pi f_2 \frac{s}{v} \right) \right]$$

$$= \phi_1 + 2\pi f_1 \frac{s}{v} - \phi_2 + 2\pi f_1 \frac{s}{v} - 2\pi f_2 \frac{s}{v} - \phi_1 + \phi_2 + 2\pi f_2 \frac{s}{v}$$

$$= 2\pi f_1 \frac{2s}{v}$$

f_2 is eliminated from the equation and the system behaves in an identical manner to one where the double path phase delay is measured for f_1 transmitted from the master via the remote and back to the master. In practice this system of utilising only one frequency would be complex due to the necessity of separating the outgoing and returned signals. The above system is a simpler approach which achieves the same nett result. In addition, the phase is measured at a low frequency of $f_1 - f_2$ (1 kHz) which enables greater instrumental accuracy to be obtained.

The specification of the MRB 201 is as follows. (Note that the MRB 201 may be used for base measurement ashore with improved accuracy. The distance reading is multiplied by the reciprocal of the refractive index assumed and then by a value for the refractive index obtained in the usual way from meteorological observations made at the time of measurement.)

Range: 1 to 50 km assuming adequate line of sight conditions.

Accuracy: under dynamic conditions, better than ± 1 m.
        under static conditions, better than ± 0.5 m ± 3 p.p.m.

Resolution: 0·1 m.

Beamwidth:

	Rotating antenna	Normal antenna on instrument
horizontal	27°	24°
vertical	22°	20°

Number of operators required: two to three depending on installation and system. Remote units can be left unattended after initial set-up.

Display: DRI with a six-digit display (Nixie type tubes) giving integrated fine readings and provision for spot checks on coarse patterns; or

DRO with vernier dial allowing each pattern to be read to three figures.

Measuring pattern frequencies:

'Master' function	'Remote' function
A 1 498 470 Hz	A -1 499 470 Hz
B 1 496 972 Hz	A $+1$ 497 470 Hz
C 1 483 485 Hz	B 1 495 972 Hz
D 1 348 623 Hz	C 1 482 485 Hz
	D 1 347 623 Hz

These frequencies are calculated to give direct read-out in metres, assuming an average refractive index of 1·000 330.

Antenna: Standard: rectangular paraboloidal reflector with dipole.
        Optional: rotating twin aerial assembly with standard reflector and dipole.

Carrier frequency: 2800–3200 MHz. An alternative frequency range of 3200–3600 MHz is available.

Carrier output power: not less than 200 mW.

Current consumption: basic unit + DRI: $3\frac{1}{2}$ amp nominal.
basic unit + DRO: $2\frac{1}{2}$ amp nominal.
(operating voltage 10·5–14 V d.c.)
Weight: basic unit: 13·68 kg.
basic unit with standard antenna and DRI: 15·68 kg.
basic unit with standard antenna and DRO: 14·7 kg.
Dimensions of basic unit closed for transport: 300 × 325 × 285 mm.
Unit function: each basic unit has dual 'master/remote' function.
Units used as 'remotes' do not require the DRI since the DRO is normally used for baseline measurements.

The Decca Trisponder 202A (Plate 2)
As the name implies, three stations are involved: a 'mobile' and two 'remotes'. Pulsed transmissions in the X-band and digital measuring techniques are employed in a range–range configuration giving the location of the mobile with respect to two surveyed points at which the small remotes are installed.

Up to four remotes can be accommodated on one system. The radio pulses from the mobile transmitter are coded to trigger a selected two of these three or four transponders which, after a fixed delay, send back answering pulses. By judicious placing of the four remotes the survey area is greatly increased.

Up to three vessels and/or helicopters can use the system together, each having a mobile aboard. When set in the auto-range mode the Trisponder provides a fix every two seconds. To do so it gathers in about 0·2 s all the data it needs and displays the ranges until the next fix. It is theoretically possible to use four further mobiles (total 5) to fit into the remaining 0·8 s, but because of fringe effects it is advisable to restrict the total of mobiles to three per system.

For its measurement of distance, just as in radar, the Trisponder relies upon the nearly constant velocity of X-band radio waves through the atmosphere and the time lapse between the emission of the pulse at the mobile station and its reception from the transponder concerned.

The two radio transmission paths are one-way streets. When the mobile emits radio pulses all remotes receive them; but the pulses are in code: each remote responds only to a pulse sent in its own code and acknowledges this by sending a pulse in return. The mobile at the same time is looking for this coded pulse and will therefore reject any other signals originating either from inside or outside the system, so minimising possible interference. If, for some reason, such as the very close proximity of radars, interference becomes excessive, the Trisponder may indicate that it has an RF link fault. By switching the mobile's distance measuring unit (DMU) to the times ten ($\times 10$) mode, the system will make repeated attempts to break through the interference with a shorter sequence of codes. When a good reading of data is decoded the distance will be indicated in 10-metre increments since the DMU will not be able to collect enough data to give a 1-metre measurement. Normally if the DMU cannot get a reading in the $\times 10$ mode the difficulty is either a malfunction in the RF link or that line-of-sight is no longer available.

The mobile station comprises:

the DMU which contains all the controls and displays for the operation, together with all the electronics for interrogation and identification, and for distance measurements;
the base unit with omni-directional antenna;
the 24-V d.c. supply.

Each remote station comprises:

the transponder with directional antenna;
the 24-volt battery supply.

The DMU controls all the Trisponder operations. All controls and read-out indicators are on the front panel. Power for the base unit is taken from the DMU via a power/signal cable. The DMU originates signals to the base unit for transmission to the remotes. The return signals from the remotes are passed from the base unit to the DMU via the same cable.

The DMU has two indicator channels, each of which actuates a set of five numeric tubes for indicating the range. The data which feed these indicators are also available at an external connector to pass the BCD signals to peripheral equipment such as a printer, paper tape punch, magnetic tape recorder or track plotter.

The base unit houses the transmitting and receiving circuitry and the antenna. The latter is an omni-directional vertical element mounted on the top of, and integral with, the unit. All range measurements to the remote are measured from the antenna. The unit is connected to the DMU via a 15·4 m cable which conveys the power supply as well as the signals.

The remote station is completely self-contained except for its power supply. Batteries are normally used. The unit can be mounted on a standard tripod or upon a pipe, or it can be fixed to any handy structure where a clear field of view is obtained. Operation is automatic and no attendance is necessary.

The standard Trisponder system can accommodate up to four remote stations. Calibration is necessary for each DMU-remote pair, to allow for the delay at the remote between reception of the mobile signal and transmission of its coded response. All adjustments for variations in these stations are carried out at the DMU by a range-calibrating control—one for each of the four remotes.

The Trisponder was designed originally to measure to an accuracy of ±3 m and this is well outside the timing inaccuracies and propagation velocity variations which may be experienced.

By virtue of the intermittent nature of the transmissions, power requirements are low and the equipment is extremely light and small.

Since the mobile employs an omni-directional antenna it is necessary only to orientate the directional antennas of the remotes roughly in the direction of the survey vessel.

The specification of the Trisponder is as follows:

Range: Up to 80 km (given line of sight conditions).
Accuracy: ±3 m.
Resolution: 1 m (fine); 10 m (coarse, or '×10 mode').
Number of operators required: none in auto-range mode, except to record ranges at DMU if required.
Display: two ranges simultaneously on 5-digit numeric tubes in DMU, manually on demand or automatically at 1-second intervals, plus BCD output for peripherals.
Antenna: omni-directional on mobile; directional type with slotted waveguides on remotes.
Carrier frequency: 9200–9500 MHz.
Carrier output power: 1 kW (max.).
Current consumption: mobile: 60 W; remotes: 30 W (all use 24 V d.c.).
Weight: DMU: 11 kg.
        Base and remotes: 7·7 kg each.
Dimensions: DMU: 406 × 305 × 216 mm.
            Base and remote units: 160 × 240 × 320 mm.
Speed of dynamic operation: up to 100 knots.

3.3.3 Medium Range Systems

It is in this category that the potential of EPF is exploited to the full. The MF frequencies (1–5 MHz) are used, propagating by ground-wave until attenuated due to the ground conductivity. Attainable ranges (which are not limited by earth's curvature) vary with this conductivity and with the output power of the transmissions, and may be from 150 to 1200 km. Though portable, the components are bulkier and heavier than those of microwave systems, and the much longer baselines warrant careful selection of station sites and the co-ordinating of aerial locations to a high standard of accuracy. Therefore the master-remote installations tend to be set up as semi-permanent *chains*. Distance measurement is invariably by phase comparison and if the remote signals are phase-locked to the master transmissions (as is often the case) they are said to be 'enslaved', and the remote station becomes a *slave*.

The master may be ship-borne, working in conjunction with two slaves to produce position circles of range centred on the slaves ashore, as in the microwave systems. This mode is termed *range-range* or *two-range*. Alternatively, all three stations may be installed ashore, the ship requiring only a receiver to detect the difference in phase between the signals of master and slaves. The ship then senses its position as the point of intersection of two hyperbolae representing lines of constant phase difference between the signals of each master-slave pair. This is the *hyperbolic* mode (Fig. 10).

Lattices may be drawn to facilitate the plotting of fixes in the manner of Fig. 10. In the two-range mode, the lattice will consist of concentric circles, centred on each shore station. The circles will represent lines of zero phase difference and will therefore be separated by a distance equal to half the measurement wavelength at the scale of the lattice. Thus a vessel at A in Fig. 10, transmitting to Slave 1, will receive the slave's signal in phase with its own transmission since it has travelled over a path length of $2\frac{1}{2} + 2\frac{1}{2} = 5$ wavelengths. The phase difference is zero and the vessel is located on the fifth position circle relative to Slave 1. These circles of zero phase difference at intervals of one half-wavelength mark the boundaries of *lanes*.* Typical equipment is capable of resolving phase difference within 0·01 lane and the precision of the system is thus related to the frequency of the transmissions.

In practice, the assumed velocity of propagation and operating frequency must be known and the projection scale factor allowed for in the construction of these lattices. (Variations in scale factor are not normally allowed for in the case of microwave systems as their short

* In the hyperbolic lattice, lanes are identified by the hyperbolae of constant phase difference which can be drawn by joining the intersections of the appropriate master and slave two-range circles (see Fig. 11).

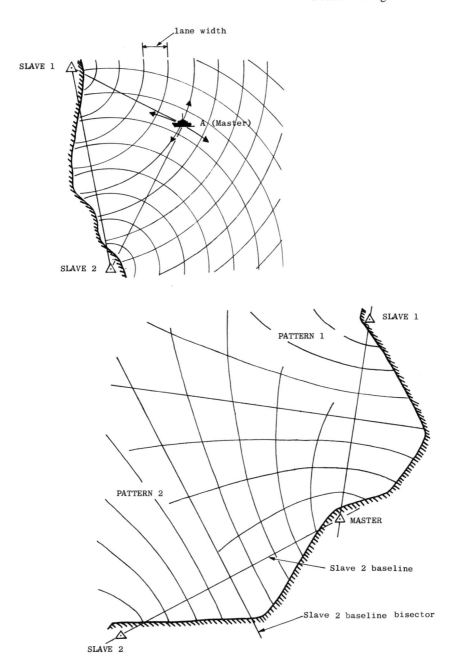

Fig. 10. The two-range (top) and hyperbolic (bottom) modes of operation

range normally enables the 'flat earth' assumption to hold good.) Calibration of the chain is essential, the vessel's position being fixed by methods which are more accurate than the system being calibrated, and the read-out is adjusted to allow for the errors detected. All of these points are fully explained in existing works, notably the *Admiralty Manual of Hydrographic Surveying*, Vol. I and the Decca publication *The Uses of Hi-Fix in Hydrography* (see Bibliography).

Bearing in mind the typical system resolution of 0·01 lane, the hyperbolic lane is clearly as narrow as the two-range lane (at the same frequency) only on the baseline. Away from the baseline the hyperbolic lane width expands with range by a factor given approximately by twice that range divided by the baseline length. The resolution of the system degrades similarly. However, any number of ships

Circles at intervals of λ produce intersecting hyperbolae $\frac{\lambda}{2}$ (1 lane) apart on the baseline. Lane width between hyperbolae = lane width on baseline x cosec γ/2 where γ is the angle subtended by the baseline at the point of observation

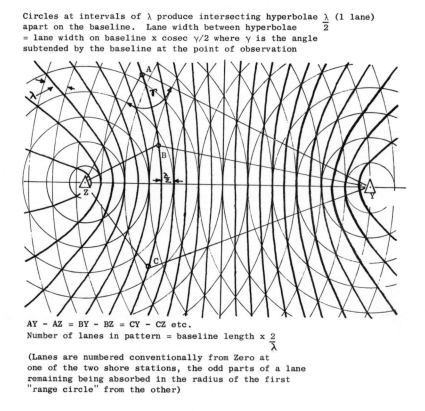

AY - AZ = BY - BZ = CY - CZ etc.
Number of lanes in pattern = baseline length x $\frac{2}{\lambda}$

(Lanes are numbered conventionally from Zero at one of the two shore stations, the odd parts of a lane remaining being absorbed in the radius of the first "range circle" from the other)

Fig. 11. The elements of the hyperbolic lattice

with the appropriate receiver can use a hyperbolic chain. In a conventional two-range chain, the master is ship-borne and that ship only can use the chain. (The lane width and resolution remain constant, of course, regardless of range.) Several modern systems—notably Raydist and Toran 'O'—achieve multi-user, two-range operation using electronic techniques only made feasible in recent years.

These systems share a problem common to all phase comparison devices, namely the inability to measure ranges greater than half the comparison wavelength directly. All systems are capable in some way of *lane integration*: that is, the summation of lane numbers as the vessel proceeds from a point of known position in terms of system co-ordinates. However, should there be a break in reception of the phase comparison signals, or the phase measurement process be interrupted for some reason, this 'lane-count' will also be interrupted and the vessel must then return to a known point for the correct lane-count to be set into the system read-out. This can clearly be a costly and time-consuming necessity. *Lane identification* is incorporated into many systems to enable a vessel's position to be obtained within a bracket of several lanes, the usual technique being to switch momentarily to a harmonic of the measurement frequency so that an identification lane is equal to several normal lanes. The fact that resolution is worse than normal is immaterial in this context.

Repeatability is a virtue of all EPF systems relying only on pattern stability and prior calibration. Once set up, a chain may be used for the repeated relocation of a position by lane readings. This feature is invaluable in many situations. For example, a dredger may use the track-plotter chart of a pre-dredging survey and, using the same positioning chain as the survey vessel, proceed directly to the areas requiring attention. Numerous other examples may be cited: positioning an oil rig over a previously surveyed site; carrying out seismic profiling operations over previously-run sounding lines; sweeping a wreck located during sounding operations; and so on. In each case the absolute position may not be of any immediate (or eventual) importance, while precision within a local framework is vital.

A number of medium range EPF systems are available to the surveyor. All use some version of the phase comparison technique, and the following notes on two such systems illustrate how the several shortcomings are minimised or circumvented in practice. Both are described in some detail in Burnside's *Electromagnetic Distance Measurement*, and the manufacturers supply excellent sales literature on request.

The Decca Navigator System

This system, with many chains in operation throughout the world, dates from 1944. The shore-based installations comprising a master and up to three phase-locked slave stations create hyperbolic lines of position which are displayed on the 'Decometer' dials of the ship-borne receiver.

In the early 1950s the master station was installed onboard a survey vessel and 'Two-Range Decca' was born, operating on the same frequencies as Decca Navigator but enabling range circles, rather than hyperbolae, to be used. A modified version incorporating lane identification is still in use. This system is called 'Lambda' (Low AMBiguity DeccA), and many government surveying vessels have adopted the practice of setting up their own 'private' Lambda chains as required for surveys out to 500 km or so from the coast stations.

A drawback of this system, which has become more serious with the increasing number of commercial users of Decca and other systems in the past twenty years, lies in the several frequencies used. Permission to transmit on the required frequencies has become very difficult to obtain, particularly off the coasts of developed nations where the common wavebands are crowded to capacity. A single operating frequency has therefore become essential and Decca Hi-Fix incorporated a time-sharing technique, whereby the master and slave stations shared a common frequency by means of interrupted continuous-wave transmissions. The development of solid-state circuitry and other advances were incorporated in the next generation Decca system, Sea-Fix, which retains the principles of Hi-Fix.

The Toran EPF Systems

The French company Sercel are manufacturers of the Toran EPF systems. Unlike the Decca systems, the first generation stations are not enslaved and the two-range mode is not used. The shore stations need no master station, rather, they are placed in pairs to provide two independent (not necessarily adjacent) baselines and hyperbolic patterns. By not sharing a common master, the shore stations may be positioned with greater flexibility to achieve optimum angle of cut of the hyperbolae over the required area.

However, the earlier Toran systems demanded several operating frequencies and suffered even more than Lambda from the difficulties of frequency allocation. The Toran 'O' system is an excellent example of the use of the latest techniques to surmount this obstacle and offers single frequency, multi-user, two-range operation. This is made possible by the development of highly stable, atomic frequency standards, sufficiently miniaturised for incorporation in portable radio equipment, and of amplification and filtering circuits capable of discerning closely similar frequencies. Several frequencies are used, in fact, but these are all within 300 Hz of the 1·6–3 MHz centre frequency and thus inside the bracket of a single frequency allocation.

The principle of distance measurement is as follows (Fig. 12). The mobile station carries only a receiver and local frequency standard. Up to three shore stations transmit a c.w. frequency F_0, where

$$F_0 = F + f$$

F being a basic frequency of between 1·6 and 3 MHz, and $f < 300$ Hz, the actual value being different for each shore station. The stability of the transmitted frequencies F_0 is maintained within 5×10^{-12} Hz.s^{-1} (or 1 part in 4×10^{17}) by means of rubidium vapour frequency standards and synthetisers.

In the mobile receiver the incoming signal is mixed with a locally generated frequency (FL) where

$$FL = F - Fi$$

Fi being a locally generated intermediate frequency.

$Fi + f$ is thus isolated and mixed with Fi to produce the characteristic incremental frequency f.

The phase of f is compared with that of an exactly similar local frequency f'. The phase difference is related to the distance (S) between shore station and mobile by the expression:

$$S = \frac{\Delta\phi \cdot V}{2\pi F_0}$$

(where $\Delta\phi$ is the phase difference and V the velocity of propagation)

$$= \frac{\Delta\phi\lambda}{2\pi}$$

(where λ is the wavelength appropriate to the frequency F_0).

A Toran 'O' chain can be used by an unlimited number of mobiles and may comprise up to six shore stations, any two or three of which are used by a mobile at a time. The read-out may be in units of distance (metres) or lanes, at will, and a BCD output can serve the usual peripherals. Note that position data consists of ranges, and not

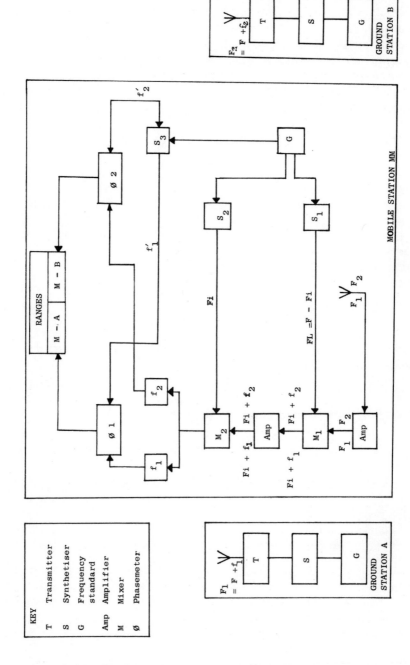

Fig. 12. Function diagram of Toran 'O'

range differences, and that the shore stations are independent and not phase locked to a master station. The frequencies transmitted are unmodulated.

As with other systems, accuracy depends on the assumed velocity of e.m. wave propagation, and chain calibration is necessary. Further, though the stability is extremely high, the frequency standards are subject to a drift of about $3 \cdot 5$ m.h^{-1}. This drift is near-linear and the position data can be corrected retrospectively. Lane identification is not provided.

The brief specification of Toran 'O' is as follows:

Range: 1000–1200 km (groundwave propagation).

Resolution: $\pm$ 1 m.

Speed of dynamic operation: (limited by Doppler shift tolerated by receiver bandwidth) 600 kts.

Carrier frequency: $1 \cdot 6$–3 MHz $\pm$ 150 Hz.

Display: 3 ranges (or lane-counts) each displayed by 7-digit numeric tubes, and BCD output to peripherals.

Dimensions and weights:

Frequency standard: 505 × 260 × 290 mm; 22 kg.

Transmitting antenna: 14 or 24 m with 15 m or 30 m radius ground mat.

Transmitter: 540 × 312 × 520 mm; 35 kg.

Mobile antenna: 4 m or 7 m whip.

Receiver: 505 × 260 × 190 mm; 20 kg + control unit: 229 × 146 × 205 mm; $1 \cdot 2$ kg + display unit: 229 × 146 × 205 mm; 2 kg.

Power requirements: 24/30 V a.c., $6 \cdot 7$–$9 \cdot 2$ A (transmitter).

Transmitter outputs: 100, 80 or 60 W.

3.3.4   Long Range Systems

The permanently established chains for use at ranges in excess of 2000 km are primarily for ocean navigation rather than surveying. Loran C and Omega are the two systems in this category, the former using pulsed signals, the latter the more familiar c.w. phase measurement technique. Both systems are used with specially latticed charts and require only a receiver in the user vessel. Loran C coverage extends over part of the globe only, chiefly the North Atlantic and North Pacific. Omega coverage is global. As Omega becomes established the less economical Loran chains may be expected to have fewer users. Further, it is possible to improve the accuracy of Omega for surveys within a particular area. For these reasons Loran C is not described here; details are given in Burnside's *Electromagnetic Distance Measurement*.

In order to discuss the surveying applications of Omega, it is necessary first to describe the system briefly. World-wide coverage is achieved using only eight shore stations transmitting in the frequency band 10–14 kHz as follows:

Station	Location	Length of transmission (s)
A	Norway	0·9
B	Trinidad	1·0
C	Hawaii, USA	1·1
D	North Dakota, USA	1·2
E	Réunion I.	1·1
F	Argentina	0·9
G	Australia	1·2
H	Japan	1·0

All transmissions are synchronised and phase-locked, the carrier frequencies being controlled by the average of three atomic frequency standards at each station, with a resulting stability of 1 μs per 10 kHz. (Omega thus provides an auxiliary facility as an extremely accurate timing device.) Time-sharing is employed with a cycle of 10 seconds, each station transmitting four frequencies: 10·2, 13·6 and 11·333 kHz, and a fourth, distinctive frequency for identification.

The signal format is shown in Fig. 13. It will be noted that the frequencies are related. The basic 10·2 kHz frequency is $\frac{3}{4}$ of the 13·6 kHz and $\frac{9}{10}$ of the 11·333 kHz, producing beat frequencies of 3·4 kHz and 1·133 kHz respectively. This provides a ready means of lane identification. The base-line lane width at 10·2 kHz is about 15 km and the user must know his position within ±7 km in order to determine his Omega lane-number. At the beat frequencies, lane widths become 45 km and 135 km respectively and lane ambiguity is reduced accordingly.

The system produces hyperbolic lines of position (LOP), and printed charts with the lattice superimposed are available to users. As with all hyperbolic systems there is lane expansion away from the baseline, but with baselines of around 8000 km this effect is minimal. Further, by judicious selection of stations the user can usually achieve the optimum angle of cut of the LOPs. The most serious factor which

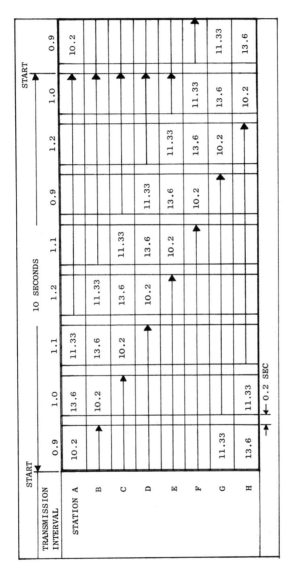

Fig. 13. The signal format of the Omega system

limits the accuracy of Omega is the varying height of the ionosphere which, with the earth's surface, forms in effect a waveguide for the VLF transmissions. Predictions of these propagation conditions are incorporated in Skywave Correction Tables and the inaccuracy of the system stems largely from the limitations of the predictions. Thus, an Omega user might expect to locate his position within ± 1 km by day and ± 2 km by night.

This absolute accuracy is inadequate for most surveying purposes. The repeatability of the system is much more useful, and Omega can be used effectively, for example, as a track-following aid between satellite fixes. Much work—on a national basis at present—is being carried out by several countries on *Differential Omega*. The theory is extremely simple and is, in fact, analogous to the use of monitor stations in medium range EPF systems. Since instrumental accuracy in the Omega system is already high, it requires only to minimise the propagation uncertainties to improve the absolute accuracy. By reading the Omega transmissions at known points onshore (i.e. monitor stations), the difference between absolute and Omega positions is obtained. This 'offset' can then be transmitted (by any means, e.g. normal ship-shore HF net) as frequently as may be desired to ships in the vicinity. The necessary correction can then be applied to the phase meters of the ship's Omega receiver so that corrected lane numbers are output direct.

Trials have shown that within 200 km of the monitor station the accuracy of Differential Omega is proportional to the vessel's distance from the monitor. Thus, at 100 km the accuracy is about 200 m, at 200 km about 400 m. Between 200 km and 500 km from the monitor an accuracy of 400–500 m may be expected, and further away than 500 km the monitor corrections become increasingly less effective.

In due time it is anticipated that international agreement may result in the establishment of monitor stations world-wide, broadcasting corrections for areas within 500 km of their location on common wavebands. The benefit to surveyors would be tremendous. Though less accurate than the medium range systems, a single Differential Omega receiver at a cost of around £3000 (see Plate 3) will enable a vessel anywhere in the world to utilise a positioning network with no setting up, calibration, maintenance or monitoring responsibilities whatever. The accuracy obtainable would be consistent, day and night and in all weather conditions. There is every indication that this happy state of affairs will reach fruition within a decade.

3.3.5 Position-fixing by Artificial Satellite

Of all available methods, the world-wide system of fixation by artificial satellite holds the promise of near-universal use in the long term. Admittedly, the satellite receiver compares in cost with the medium range EPF chain, the necessary computer is an expensive substitute for the human brain in the processing of data to obtain a fix, and at present fixes are possible only at intervals of up to two hours or so. However, the bracket between satellite and conventional methods is closing inexorably. The absolute accuracy obtainable ashore and afloat is high and improves steadily as knowledge of the earth's gravity and the satellites' orbital parameters grows; more and more 'conventional' systems are being incorporated into automated systems with the computer at the hub; electronics technology is rapidly bringing down both the cost and bulk of the computer; and the savings in manpower made possible show returns in terms of cost, vessel size and logistics. The necessary track control between fixes may be performed by a variety of options such as Doppler sonar, inertial systems and, of course, Omega (and other phase comparison systems) where the repeatability becomes an invaluable quality.

Though the close and short range methods are unlikely to be superseded, shore control by satellite could well be the norm and, for all survey work of magnitude in such areas as hydrocarbons exploration and the charting of deep draught tanker routes, automated systems based on satellite fixation are more than probable.

The US Navy Navigation System TRANSIT has been accepted as the satellite position-fixing method most suited to the surveyor's needs ashore and afloat. The Doppler principles employed are simple. When a receiver and transmitter approach each other the received frequency is higher than the transmitted frequency by an amount which is directly proportional to the relative velocity of the transmitter and receiver. The received frequency is *lower* than that transmitted when the receiver and transmitter are moving apart. Integration of the velocity data obtained with respect to time gives the change of distance between transmitter and receiver (see also p. 50). The system's geodetic applications are outside the scope of this book. These notes are therefore confined chiefly to the dynamic positioning and other facets pertinent to the marine applications.

Very briefly, TRANSIT operates as follows: a number of satellites in polar orbit are tracked by accurately co-ordinated stations around the world. The Doppler frequency shifts of the satellite transmissions are recorded as a function of time by the tracking stations and transmitted to a computing centre in the USA. From these data a navigation message is computed and supplied to an injection station for transmission to the satellite concerned. The satellite's memory receives an up-dated navigation message at approximately 12-hourly intervals and the satellite emits its message continuously, each transmission taking exactly two minutes. The satellite transmissions are made simultaneously at frequencies of (very nearly) 150 and 400 MHz and consist of two parts: the first is fixed and describes a smooth elliptical orbit, the second consists of corrections which modify the perfect ellipse to the actual satellite position at eight 2-minute time marks.

A vessel having an appropriate receiver and associated computer receives the navigation message and measures the Doppler frequency shift on both frequencies as the satellite rises, crosses the closest point of approach (CPA), and sets. These data are passed to the computer, where the changes in slant range of the satellite at successive time marks are computed from the integrated Doppler data. The computer compares these 'measured' slant ranges with theoretical values obtained from the known satellite positions and the ship's position as estimated by the user. By a process of iteration, modified estimates are made of the ship's position until agreement with the 'measured' values is reached. The finally accepted latitude and longitude are then printed out by the computer or, in an integrated system, passed to the interface for up-dating of the track-following system and of the running track-plot.

The generalised notes above apply equally, of course, to a satellite system either ashore or afloat. In conventional geodetic work ashore, many repetitions of observations are made to reduce the statistical errors to the minimum. Similarly, by observing many passes of all possible satellites over a period of, say, two weeks, accuracy in the absolute positioning of a point onshore of 5 m may be achieved.

The errors in measurements both ashore and afloat may be due to:

(i)   noise and instability in receiver circuits;
(ii) inaccurate estimate of antenna height above the surface of the gravity model (geoid) used in the satellite's navigational message;
(iii) propagation anomalies;
(iv) errors in the ephemeral predictions contained in the satellite navigation message.

Instrumental errors may account for some 10 m or so in the derived position, noise being a greatly variable element and the stability of the local reference oscillators being around 1 part in 10^{11}. On the whole, however, provided satellite elevations are reasonable and given average reception conditions, such errors as may accrue will be virtually negligible.

Sea-borne receivers are much less prone to error due to uncertainties in antenna height, of course, mean sea level being virtually the surface of the geoid. Because of the N–S orbital path of the satellites, errors will largely be those in longitude, and will vary considerably with satellite elevation.

Propagation is affected by tropospheric and ionospheric refraction. The former varies with signal frequency, as does the Doppler shift, and is impossible to isolate. The latter varies inversely as frequency and is measured and allowed for by the dual-frequency transmission. (The ionospheric refraction effect on the Doppler count of the 150 MHz signal is 8/3 of that experienced by the 400 MHz count.)

The ephemeral predictions can only be as accurate as the geoid in use, and this may be expected to improve steadily as knowledge of the earth's gravity is accumulated.

At sea there is inevitable further loss of accuracy. A vessel underway has a certain velocity in relation to the earth's rotation. Since measurement by Doppler principles relies wholly on knowledge of the relative velocities between transmitter and receiver it follows that uncertainties in ship velocity are the major sources of error. In a position fix at sea, the computation has to define the ship's *path* which best fits the Doppler-measured range differences during the satellite pass, rather than the single position of a stationary observer.

Again due to the satellites' N–S orbital path, errors in the northerly component of ship velocity have the more serious effect on the accuracy of the Doppler counts. It therefore follows that the track-following method, which provides the ship speed and heading data, has an important bearing on the overall positioning accuracy obtainable. In this respect, inertial systems are quite superior, and Doppler sonar systems in continental shelf depth, though inferior to inertial systems, are far better than other methods such as gyro compass and ship's log.

Another element contributing to positioning accuracy is the software of the system (the program) which controls the efficiency with which the Doppler range data are used to compute latitude and longitude at each time mark, and thus the 'best fit' path of the vessel during the satellite pass. A recent improvement perfected by the Magnavox Company and incorporated in their MX/702/hp system is in the use of 'short Doppler counts'. The program is designed to perform five 23-second counts during a two-minute navigational message, a total of 40 counts being possible in one satellite pass (as distinct from the usual 8 or 9). The computation includes for inter-polation of the satellite position between the even-minute time marks, and as a result a reliable fix can usually be obtained even when parts of the satellite message are distorted by noise.

In conclusion, it should be borne in mind that the state of the art makes possible offshore fixes with an absolute accuracy of around 50 m, given optimum conditions and good ship-velocity data. This currently places the satellite system slightly behind the medium range hyperbolic EPF systems (in optimum conditions) and far ahead of the long range EPF systems. Thus, the satellite system can be said to be justified:

(a) whenever a vessel requires to operate with maximum accuracy at distances offshore greater than, say, 300 km;
(b) whenever a computer is required as a component of an auto-mated data acquisition or processing system;
(c) whenever a vessel requires maximum positioning accuracy in areas where the use of medium range EPF is impracticable (e.g. due to difficulties of frequency allocations, siting of shore stations, extent of area of operations, etc);
(d) (if already installed) as a complement to EPF systems for lane identification and calibration, and, conversely, by using the EPF system for track-following between satellite fixes.
A final word of warning: the surveyor should ensure that the program allows for conversion of the data to geographical position on the projection and grid in use.

A satellite position-fixing system, the Magnavox 702A, is illustrated in Plate 4.

3.4   Track Control

Many methods are employed to determine the survey vessel's progress along its intended track.

If a continuous read-out of position is obtained by, for example, an EPF system forming part of an automated data-processing and plotting system, a fix may be plotted at intervals of a few seconds and the track shown as a series of dots representing position at discrete intervals of time. As already mentioned, the same system may be used to indicate to the helmsman the direction he must steer to maintain the intended track. The system might even control the rudder so that the track is steered automatically, and the vessel is then directed and fixed in a predetermined manner as other data (hydrographic, geophysical, etc.) are acquired.

In a conventional, close range survey with visual or microwave position-fixing, position may be fixed once every 30 or 60 s and the vessel must be assumed to occupy, between fixes, positions represented by the straight line joining successive fixes. The helmsman will then adopt some method of maintaining the line so that the straight line truly portrays the track followed.

Much depends in this context on the scale of the survey. The pencil line joining fixes, perhaps 0·2 mm wide, represents 10 m on a scale of 1:50 000 but only 1 m on a scale of 1:5000. This indicates the theoretical maximum amount by which the vessel should be allowed to deviate from the line. According to the precision dictated by the scale of the survey, therefore, and the instrumentation available, the surveyor might adopt any of the following expedients:

(i) Steering a compass course, estimating the offset required to counteract drift due to wind and current.

(ii) Steering along a transit, i.e. keeping two marks in line.

(iii) Steering as directed by a sextant, theodolite or laser observer ashore, his instrument pointing along the intended direction of the sounding line and communicating by flags, light or radio with the survey vessel.

(iv) Steering along an arc of constant subtended sextant angle, or range, or along a 'lane' boundary of an EPF system lattice.

Where the interval between fixes is too long to guarantee the vessel's adherence to the straight line between fixes, as with satellite fixing, for example, additional instrumentation is necessary to assess the course and speed made good by the vessel. Examples are the inertial and Doppler sonar systems of track-following.

Inertial systems employ gyros and accelerometers to sense the vessel's motion across the earth's surface. Their accuracy is entirely dependent on the precision with which their component parts are engineered. They are therefore highly expensive and unlikely to find favour outside their present area of use in government service.

Doppler sonar is, in depths less than 200 m, a much less costly and quite effective alternative, giving the vessel's true velocity and heading. Beyond this depth the transmitted pulses are returned on reflection from water layers rather than the seabed and the resulting velocity and distance data are relative to the water mass. In these circumstances Doppler sonar is little better than an efficient ship's log. (It should be stressed that only distance and direction relative to the earth's surface as distinct from the moving water mass is of any surveying value.)

The principles of acoustic instrumentation are included in Chapter 4, and the following description is limited to non-technical detail (see Fig. 14).

The Doppler frequency shift due to the relative motion of a vessel over the seabed may be measured in an acoustic pulse transmitted from the vessel towards the seabed and received after reflection on

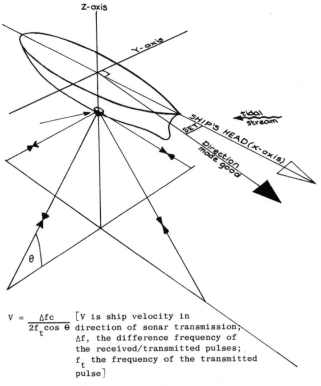

$$V = \frac{\Delta f c}{2 f_t \cos \theta}$$

[V is ship velocity in direction of sonar transmission; Δf, the difference frequency of the received/transmitted pulses; f_t the frequency of the transmitted pulse]

Fig. 14. The basic principles of Doppler sonar track-following

return to the vessel. The Doppler data may be resolved into acceleration, integrated with respect to time to produce velocity and again integrated to produce distance. The angle of the transmission to the horizontal must be known, and the velocity of propagation of the acoustic pulse. The data will give information of velocity and distance along the line of the transmission—usually along the heading of the vessel. The vessel may be moving crab-wise over the seabed due to the influence of tidal stream or current ('set'), and the vessel will also move in pitch and roll due to wave action. Thus, if the ship's head forms an X-axis and the line at right angles (from port to starboard) the Y-axis, and if four pulses are transmitted—ahead, astern, to port and to starboard—the X and Y components of the ship's velocity may be measured and the effects of roll and pitch largely compensated. If the ship's heading data (by gyro compass) are then applied to the X–Y data, the ship's true 'course and speed made good' over the seabed may be calculated. The simpler Doppler systems, most usually employed for docking large vessels, give only velocity forward or astern and velocity port or starboard. A surveying system suitable for use in a satellite positioning system will resolve for the true course and speed.

CHAPTER FOUR

Acoustic Measurement and Investigation

4.1 Introduction

In this chapter it is intended to discuss the use of acoustic waves for performing measurement and investigation functions in the seawater environment.

Sonar is the acronym from 'SOund NAvigation and Ranging' which is given to instruments which record information from underwater transmissions analogous to Radar on the surface. In this category are included the echo-sounder, the side-scan sonar, searchlight sonar, sector-scan sonar and Doppler sonar.

Doppler sonar is described in the previous chapter. The echo-sounder and side-scan sonar are the surveyor's major source of data and are discussed in some detail. *Searchlight* sonar is the informal term given to the rotating transmitter with cone-shaped beam used primarily for the detection of submarines (originally 'Asdic') and in a peacetime role in fish detection. Its surveying uses are restricted chiefly to government agencies and in any case the side-scan sonar performs the same wreck location and bottom search functions with greater efficiency; therefore searchlight sonars are not discussed further here. Some notes on sector-scan sonar are given, since this instrument is indicative of developments which will probably involve the surveyor in the foreseeable future.

Other types of acoustic instrumentation are varied and increasing. Perhaps of most direct interest to the surveyor are acoustic beacons, for positioning control and for the marking of instruments on the seabed or as a means of checking the lane-count of EPF systems in areas remote from shore.

4.2   Sonar Instrumentation

4.2.1   The ideal means of carrying out measurement and investigation through the opaque waters of the sea might appear to be, as on land, by electromagnetic instruments such as the Tellurometer and Radar. Electromagnetic waves are very quickly dispersed and attenuated in seawater, however, only VLF waves having any effective penetration. Instead, the same principles can be applied using pressure waves, and for radio waves we substitute sound (or ultra-sonic) waves for the transmission of energy through the water mass.

In the following brief outline of acoustic wave propagation in seawater, the land surveyor will constantly note the parallels with e.m. wave propagation in air and the associated instrumentation.

4.2.2   Transducers and Beam Shaping

Pressure waves in the frequency range 1–300 kHz are the most commonly used. They are created by a vibrating surface, its resonant frequency being that of the transmission frequency required. Usually, the vibrating surface is 'driven' by applying electrical power to a ceramic 'motor', or direct to the surface, which then converts the electrical power into sound power by vibrating in contact with the seawater medium. Such a system is a *transducer*. A transducer which is caused to vibrate by sound pressure waves will similarly convert the sound power into electrical power.* Therefore a transducer can transmit sound power through the seawater medium and receive the sound power reflected by a target in the path of the transmission in a manner exactly analogous to that in which Radar functions with e.m. waves. Transducers acting as transmitters are sometimes called *projectors*; those acting as receivers, *hydrophones*. Transducers for surveying are designed to produce *beams* of sound power of various shapes, normally concentrating the energy along an axis orthogonal to the radiating surface.† The determining factor is the size of the

---

* Transducers do not necessarily react only to a resonant frequency. Some, such as hydrophones, have a 'flat' response to a range of frequencies.

† Transducers can also be 'steered', i.e. the output energy may be concentrated in directions other than the axis, or curved to insonify a large volume of seawater.

surface in wavelengths at the resonant (transmitted) frequency. Thus, a long, narrow surface will produce a narrow beam in the plane orthogonal to the long dimension. A circular transducer will produce a cone-shaped beam (see Fig. 15).

Beamwidth indicates the degree of *directivity* achieved—that is, the ability of a transducer to concentrate the sound power. Very roughly, the beamwidth (β^0) of a circular transducer is given by the expression

$$\beta^0 = \frac{65\lambda}{d}$$

where λ is the wavelength of the transmitted or received frequency and d the diameter of the radiating surface (both in the same units), and that of a rectangular transducer by the expression

$$\beta^0 = \frac{50\lambda}{L}$$

where L is the dimension of the radiating surface orthogonal to the plane of the beam being considered.

An average value for the velocity of propagation of acoustic waves in seawater (*sound velocity*) is 1500 m.s^{-1}. An acoustic transmission of 30 kHz will therefore have a wavelength of about (1500 ÷ 30 000) m, or 50 mm. For a conical beamwidth of 30°, therefore, a transducer of diameter (65 × 0·05 ÷ 30) m, or about 110 mm, is required. The beamwidth of a rectangular transducer of dimensions 1 m × 0·1 m will be (50 × 0·05 ÷ 1) degrees in one plane and (50 × 0·05 ÷ 0·1) in the other, or 2·5° × 25°.

The beamwidth is defined as the angle between the points at which the sound energy intensity has fallen to half that along the beam centre axis. Therefore a beam is described, for example, as 'Thirty degrees between half-power points' or, using the decibel scale, as 'Thirty degrees between −3 dB points'.*

---

* The sound intensity at the limit of the beam as defined (say, I_1) is half the intensity at the centre (say, I_2). Thus

$$I_1 = \frac{I_2}{2}$$

and in the decibel notation the intensity ratio is

$$10 \, \mathrm{Log}_{10} \frac{I_2}{I_1} \quad \text{or} \quad 10 \, \mathrm{Log}_{10} 2 = 3 \, \mathrm{dB}$$

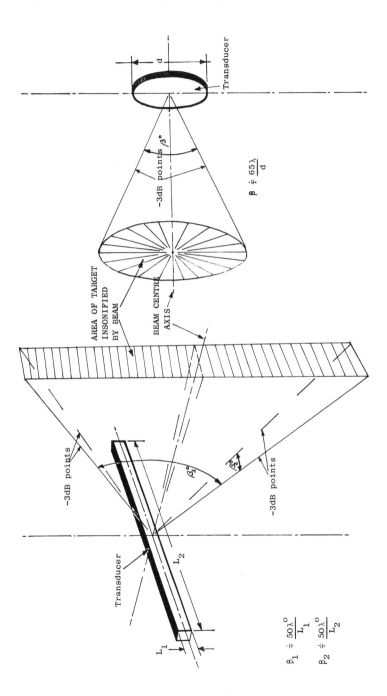

Fig. 15. The beam-shaping characteristics of (left) a rectangular and (right) a circular transducer

4.2.3 Frequency

Clearly, the wavelength (and thus, frequency) of the transmission is of great importance. For accurate measurement to a point (say, on the seabed, as is the case with depth measurement) a high resolution, narrow beam is required. A further characteristic of acoustic wave propagation is that the higher the frequency the shorter is the wavelength and the narrower the beamwidth for a given size of transducer. To use the previous example where the transducer size is 1 m × 0·1 m: if the frequency is raised from 30 kHz to 300 kHz, the wavelength becomes (1500 ÷ 300 000) m or 5 mm and the beamwidth is reduced to 0·25° × 2·5°. If a long range echo-sounder is required for deep water sounding, the required beamwidth might be specified as 5°. But in this case, the transducer would need to have a diameter of 650 mm, and for a 2° beam, 1·6 m. Such a transducer manufactured from nickel-alloy plates or similar will be heavy, bulky and expensive, and in practice a compromise is reached whereby some resolution of the beam is sacrificed by accepting a beamwidth of, say, 30°, with corresponding reduction in the transducer size (to 108 mm) and cost. (In any case the beam will be subjected to wave and swell motion in the same way as is the survey vessel and the wider beam will cater for some rolling and pitching. The alternative would be to stabilise the beam in the vertical by gyros and servo motors or some similar expedient. This is invariably prohibited by cost limitations except for major government research and like projects.)

4.2.4 The Acoustic Transmission

Pulsed transmissions are almost invariably used. In simple echo-sounders the pulse may be created by applying an electrical discharge momentarily to the transducer, causing it to contract, or otherwise undergo a physical change. The transducer diaphragm then vibrates at its natural frequency as it reverts to its original state. The resulting pulse of acoustic power has an initial high amplitude which tails off rapidly as the vibrations diminish. In more refined instruments, particularly echo-sounders for deep waters (precision depth recorders —PDRs) and side-scan sonar equipment, the pulse is caused by the gating of electrical power from a continuously oscillating source. Such an oscillator is often crystal controlled, the resulting frequency is quite pure and the outgoing pulse shape is contained within an almost square 'envelope'.

4.2.5 Resolution

The resolution of an acoustic equipment can define either its measuring precision or detection capabilities and is determined by a number of factors, including:

(a) the pulse duration or length;
(b) the angle of incidence of the acoustic wave-front to the target;
(c) the sensitivity and resolution of the recording medium;
(d) the nature of the target;
(e) the beamwidth of the transmission.

A gated pulse has a finite length determined by the frequency, the velocity of propagation and the duration of the pulse. For example, a pulse lasting 1 ms at a frequency of 15 kHz will consist of a 'bundle' of 15 cycles. If the sound velocity is $1500 \text{ m} . \text{s}^{-1}$, the wavelength is $0 \cdot 1$ m and 15 cycles will produce a pulse length of $1 \cdot 5$ metres. If two objects up to half the pulse length apart in range lie in the path of the acoustic ray, they will reflect, and be recorded as, a single target. If they are further apart than this, they will be recorded as two separate *echoes*. The resolution is then said to be $0 \cdot 75$ m, or half the pulse length. If the ray path is not incident to the normal to the target, however, the effective pulse length will be greater and the resolution will be degraded.

The above points and certain standard terms used in this context are illustrated in Fig. 16.

There are limits, of course, to the minimum possible pulse length, and in any case the display medium may not be capable of resolving the various reflected echoes received by the receiver circuitry. Electrical styli normally record on electro-chemical or carbon-backed paper and the stylus marking may well represent an echo of greater extent on the range scale than the length of the pulse. Recording media such as magnetic tape do not suffer from this shortcoming, but nor do they provide a real-time portrayal of the data acquired.

Targets may be of widely varying quality. In the sonar sense this is expressed as reflectivity or target strength. The acoustic power returned by a target will depend upon the density of its constituent materials, upon its size and inclination to the transducer. For example,

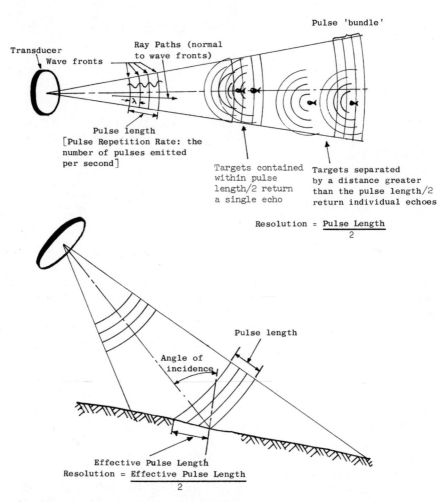

Fig. 16. The effects of pulse length and angle of incidence on resolution

air is of a low density relative to seawater and therefore affords good reflectivity. A target such as the seabed or a shipwreck will return more power to the transducer from those parts of it at right angles to the ray paths than from parts which will reflect the power away from the transducer. These factors are most important in the qualitative interpretation of a sonar record (or *trace*), an experienced observer being able to differentiate between the characteristic traces of different seabed materials, spurious echoes (from fish, weed, marine organisms), etc.

The effect of beamwidth on resolution has been noted earlier. In depth measurement, the point of interest is naturally that lying vertically beneath the transducer. However, the echo-sounder records the earliest return from its transmission, i.e. the echo having travelled the shortest distance. In the case of a wide beam, this may well be from some target within the insonified area of the beam not directly beneath the transducer. In some cases, particularly in deep water where the insonified area may be considerable, a trace is recorded which bears very little resemblance to the true configuration of the seabed (see Fig. 17).

Over a steadily sloping seabed, the point of interest is always offset from the point of measurement (see Fig. 18).

4.2.6 Accuracy

Throughout the preceding paragraphs a velocity of propagation of the acoustic waves in seawater has been assumed. The actual value is the surveyor's most serious uncertainty and source of error in measurement underwater. The factors which affect this sound velocity similarly influence the behaviour of the beam, and the combined consequences merit careful consideration.

The velocity may be measured with relatively simple instrumentation (sound velocity meters) at any depth, but these point-observations do not satisfy the need to determine the *average* velocity over the entire path between transducer and target—a distance which varies continuously along a sounding line.

Sound velocity is a function of temperature, salinity and density (in descending order of magnitude). All three properties change diurnally, seasonally and with random or periodic influences such as the tidal stream, rainfall, etc. All three vary with depth also, the very parameter we most frequently require to measure! A full study of sound velocity takes the student deep into physics and physical oceanography and cannot appropriately be included here. Several methods may be used to cope with the problem, none of which is entirely satisfactory, and it must be stated at the outset that accuracies in acoustic measurement in seawater of better than one part in 200 should not be expected unless resorting to time-consuming and costly observations.

The most common calibration technique for use in depths to about 30 m is termed the *bar check*. As the name implies, a bar (or disc) is set horizontally beneath the transducer on marked lines at various depths,

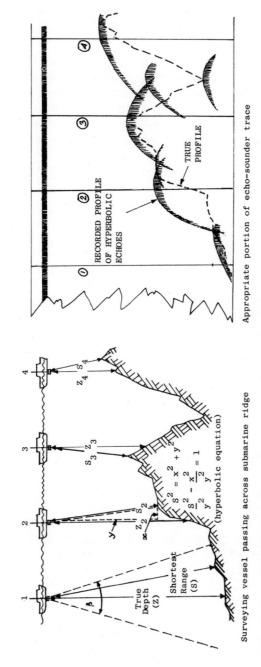

Fig. 17. The effect of beamwidth on the echo-sounder record. Note: the shortest path for each pulse may be to points to port, starboard, or astern of the vessel as well as ahead as shown

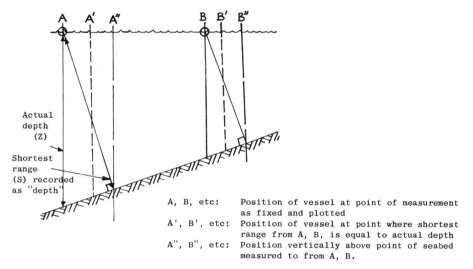

Fig. 18. Sounding over a steady gradient

and the echo-sounder recorder is adjusted so that the echoes from the bar appear on the trace at the correct depths determined by the marks on the lowering lines. This technique is described in detail in para. 4.3.

A shortcoming of the bar check lies in the possible slope of the lowering lines leading to imprecise setting of the bar, particularly at maximum depths. Changes in sound velocity may be expected to occur with time during a day's sounding and, indeed, from place to place over the survey area. All methods are subject to this disadvantage, however, and the only solution is to recalibrate at intervals (say, twice daily or before and after a high or low water, or after heavy rainfall).

Some echo-sounder manufacturers provide a set of calibration curves for given values of temperature and salinity, which give a motor speed setting to equate the echo-sounder depth scale with the sound velocity. Field measurements of salinity and temperature at the time of calibration are implicit in this method.

As an alternative to the velocimeter method, an echo-sounder transducer may be installed in a rigid cradle with a small target such as a metal plate set at a known distance (say, 1 m) below the transducer. This rig is lowered on a marked depth-line and the survey echo-sounder is adjusted to give a read-out of 1 m at all depth settings through the water column. The average sound velocity is thus allowed for automatically.

Precision depth recorders and most sonar instrumentation other than echo-sounders incorporate a very precisely fixed motor speed, usually equating the depth/range scale with a sound velocity of 1500 m.s^{-1}. Calibration is therefore not possible in the usual way and tables are used from which corrections are obtained to allow for predicted sound velocity by region for all the world's sea areas. The corrections are applied at the draughting stage, of course. The procedure is cumbersome, and the corrections are necessarily unreliable because of the very low density of temperature and salinity observations on which the sound velocity formulae are based.

Possibly the best method is to use an instrument which gives a direct, continuous read-out of temperature, salinity and sound velocity against depth as the probe containing all the necessary sensors is lowered through the water column. A BCD output can be applied to a computer, suitably programmed to integrate and average out the readings to determine the average sound velocity. This operation need take only as long as may be required to lower and hoist the probe—in continental shelf depths a matter of perhaps 20 minutes. However, the capital cost of the necessary instrumentation equates with that of an EPF chain or a satellite receiver, and the less refined methods are normally preferred.

A useful method of checking the echo-sounder depth in the deep ocean is by the 'thermometric depth', calculated from the temperatures obtained by oceanographical thermometers.*

In waters of continental shelf depths (to 200 m or so) a simple sound velocity meter may be used and the readings obtained at intervals through the water column are then meaned arithmetically to derive average sound velocity without appreciable error. This instrument costs under £1000 and usually employs the 'sing around' principle. A transducer emits a sound pulse across a very short, accurately-fixed path length. The pulse is reflected back to the transducer which is designed to emit a second pulse on receipt of the first.

* Thermometric depth

$$Z = \frac{(T_u - T_w)}{Q \cdot \rho_m}$$

where T_u is the corrected temperature at the depth of observation as read from a thermometer which is subject to hydrostatic pressure, T_w is as T_u, but read from a thermometer which is protected from hydrostatic pressure, Q is a pressure coefficient for the unprotected thermometer and ρ_m is the mean density of the water column above depth Z.

Oceanographical tables (for ρ_m) and the calibration documents for the individual thermometers are essential for the application of the above formula. The value for Z should be accurate to about ±5 m.

Thus, the pulse repetition rate is related directly to the sound velocity, enabling a direct read-out in the required units (usually metres per second) at the deck unit.

The variables affecting the sound velocity also cause the bending of any ray which does not pass perpendicularly through the temperature—salinity—density layers of the water column. The raybending effect is negligible in echo-sounding due to the vertical pointing of the beam. In other sonar systems and in ranging from seabed acoustic beacons, etc., the accuracy may be severely affected and results misleading. The phenomenon is illustrated in Fig. 19.

4.3 The Sonar Record

4.3.1 The chief data requirements of the user of sonar instruments are quantitative—i.e. measurements of range—and qualitative, to enable the user to make an assessment of the nature of the target. The only source of data for both requirements is acoustic transmissions, and the methods employed to present the information to him must be designed to process these data accordingly.

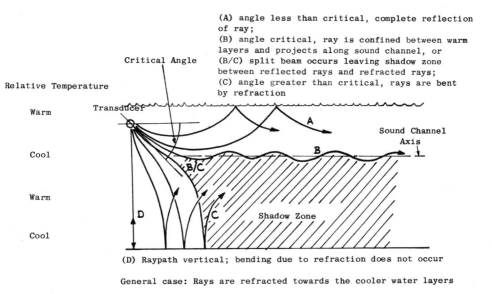

(A) angle less than critical, complete reflection of ray;
(B) angle critical, ray is confined between warm layers and projects along sound channel, or
(B/C) split beam occurs leaving shadow zone between reflected rays and refracted rays;
(C) angle greater than critical, rays are bent by refraction

(D) Raypath vertical; bending due to refraction does not occur

General case: Rays are refracted towards the cooler water layers

Fig. 19. The bending of sonar ray paths caused by the temperature structure of the water mass

For the ranging requirement the sonar instrument must have:

(a) a means of emitting the acoustic pulse;
(b) a means of receiving the returning echo pulse;
(c) a means whereby the interval of time between (a) and (b) may be measured and converted from time to distance.

For an investigation, an impression of the target shape, size and orientation must be built up from the accumulated ranging data.

For both requirements, the surveyor requires a permanent record of the results obtained—the sonar record.

All sonar equipment requires the following fundamental components:

(i)   a time base, controlling the emission of the outgoing pulse, and measuring the lapse of time before the return of the echo pulse;
(ii) a source of electrical power to activate the transmitting transducer;
(iii) transmitting and receiving transducers (transmission and reception are sometimes performed by one transducer);
(iv) a means of amplifying the weak electrical signal produced by the receiving transducer on the arrival of the echo pulse;
(v)   a means of recording the transmission and arrival times of the acoustic pulse after conversion of the time interval to distance.

4.3.2   The Echo-sounder Record

The ways in which the results are recorded vary. A typical echo-sounder is described here to illustrate the steps involved (see Fig. 20).

An electric motor drives (via a gearbox) a rotating stylus. At the point where the stylus passes over the zero of the range-scale a cam arrangement causes the closing of the transmission contacts, allowing a current to pass through the stylus and recording chart to the earthing plate (*platen*). A mark is thus made on the recording chart, or trace. The stylus rotates across the trace range-scale until the echo pulse generates a current which is applied to the stylus, and a further mark is made. The stylus next reaches the transmission point once more and the cycle is repeated. At the same time, the motor/gearbox drive is caused to move the trace in a plane at right-angles to that of the stylus movement, and the resulting succession of marks made by the stylus constitutes a time-time graph.

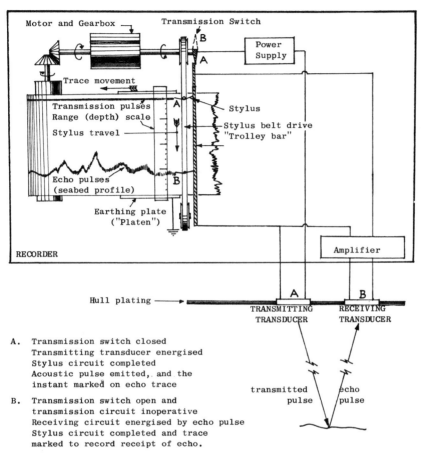

A. Transmission switch closed
Transmitting transducer energised
Stylus circuit completed
Acoustic pulse emitted, and the
instant marked on echo trace

B. Transmission switch open and
transmission circuit inoperative
Receiving circuit energised by echo pulse
Stylus circuit completed and trace
marked to record receipt of echo.

The reading of depth (z) on the scale is given by the expression

$z = \dfrac{vt}{2}$, where v is the sound velocity and t is the interval
between instants A and B, corrected for the depth below
the water level of the transducers.

Fig. 20. The sequence of events in the sounding cycle (schematic) showing the parts
of the recorder

It should be stressed that, while the stylus speed is adjusted in calibration to equate with the speed with which the acoustic pulse travels to the seabed and back, the chart speed is simply a geared-down proportion of the motor speed and does not equate with the vessel's speed except coincidentally. The 'time-time' graph thus does *not* represent a vertical-horizontal distance profile and the resulting record is *not* a true profile of the seabed. The chart will continue to move away from the stylus as long as the instrument is switched on, regardless of whether the vessel is underway or stopped.

It will be noted from Fig. 20 that one transmission mark and one depth mark provide but scant information. Indeed, the depth mark could be confused with other spots which inevitably appear on the trace due to pitting of electrical contacts, small midwater targets and the like. On the other hand, the attenuation of the transmitted acoustic power is such that, after the journey to the seabed and back, the echo pulse might be insufficiently strong for the receiver circuitry to discriminate its arrival from the background noise. The succession of stylus marks, however, produces a recognisable profile of the seabed. In addition to enabling the surveyor to appreciate the trends of the seabed topography, this profile provides 'echo correlation' and thereby raises the *detection threshold*. In other words, the chances of detection of a series of echo pulses are enhanced over the chances of detecting a single, isolated echo of the same strength.

Before the surveyor can use the echo-sounder trace, the depth measurements recorded thereon must be related to the position of the vessel to complete the necessary x-y-z ordinates of the point being surveyed. An overriding 'event mark' facility is provided for this, whereby the stylus circuit is completed (manually or automatically) throughout its travel at the point of fix. Other marks may be made, by mechanical cam or slip ring devices, such as time marks, depth scale graduations, motor speed calibration lines, and so on.

Methods of calibration are described in para. 4.2.6. The adjustments made must allow for two possible error sources—index (or transmission) error and, of course, motor speed error. The index error is constant and is simply the effect of the transmission emanating from the transducer a short distance beneath the water line. It is corrected by setting the bar check to a depth of about 2 m and adjusting the transmission setting until the depth of the echo of the bar agrees with the known depth set by the lowering lines. Motor speed error is due to its mismatching with the sound velocity and increases linearly with time. The error is therefore most marked at maximum depth. The bar is lowered to the maximum extent and the correct depth of the echo of the bar is set by adjusting the motor speed. (When a sound velocity meter is used, the motor speed equivalent to the sound velocity must be known and this can then be set by tachometer. The index error must still be removed, of course.)

An echo-trace showing a bar check is illustrated in Fig. 21.

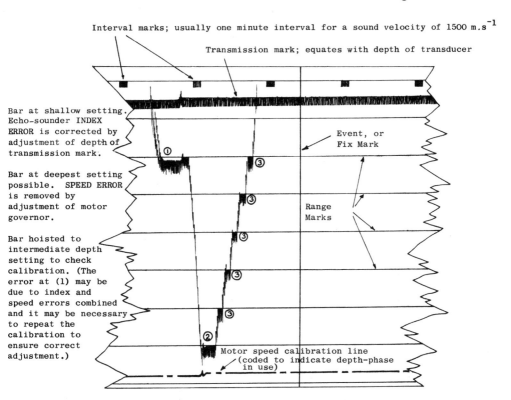

Interval marks; usually one minute interval for a sound velocity of 1500 m.s^{-1}

Transmission mark; equates with depth of transducer

Bar at shallow setting. Echo-sounder INDEX ERROR is corrected by adjustment of depth of transmission mark.

Bar at deepest setting possible. SPEED ERROR is removed by adjustment of motor governor.

Bar hoisted to intermediate depth setting to check calibration. (The error at (1) may be due to index and speed errors combined and it may be necessary to repeat the calibration to ensure correct adjustment.)

Event, or Fix Mark

Range Marks

Motor speed calibration line (coded to indicate depth-phase in use)

Note: Not all types of echo-sounder incorporate all features shown, and some employ different methods of indicating phase, range, scale zero, etc. The portion of trace illustrated is <u>linear</u>, being produced by a belt-pulley stylus drive as in Fig. 20. Plate 6 includes a <u>radial</u> trace, produced by a stylus mounted on a rotating arm.

Fig. 21. A typical bar check, and the echo-sounder trace explained

The width of the trace is usually designed to represent a number of metres depth, say 30, 50 or 100 m. With a trace representing 0–100 m, an echo from a depth of over 100 m will not be recorded, but if sufficient power had reached the receiving transducer to generate an electrical pulse, it would be available at the stylus after the appropriate time interval whether the stylus was over the trace or not. (In Fig. 20, the stylus would be on its upward path behind the trace.) However, if the transmission contacts could be moved back a distance equivalent to 100 m, the stylus would be over the trace at the instant when the echo pulse was received, and the depth could be read as 100 m plus the reading on the depth scale. The transmission 'mark' would not be recorded, of course.

This facility is termed *phasing* and may be a purely mechanical operation performed manually, or an electrical switching circuit. In some instruments the stylus is made to mark the foot of the trace with a coded line of dots and dashes to record the phase in use automatically.

The stylus will mark the trace at all times when a current is passed through it. Thus, if a fish or other object lies in the path of the pulse an echo strong enough to generate such a current may be returned. Very often, the side lobes of the beam will pick up echoes returned from shoals lying outside the main part of the beam. These echoes are called *side echoes* and can easily lead to misinterpretation of the trace (Plate 6). Similarly, the stylus will continue marking for as long as the main seabed echo is being received—possibly several milliseconds— and these returns show on the trace as a 'tail' to the leading edge of the echo.* When the pulse reaches the seabed, the major part of the acoustic power is reflected back to the surface. However, a certain amount penetrates the seabed and, if the pulse carried sufficient power initially, echoes may well be received from firm boundaries— between sediment and bedrock or between different layers of sediment—beneath the seabed. Low frequency waves suffer less from attenuation than high frequency waves and high powered pulses of sound energy at frequencies of 1–15 kHz are widely used by geophysicists to determine the geological formations of the seabed. By the same token, a surveyor's 30 kHz equipment might penetrate silty sediments and show as the seafloor the consolidated matter beneath, while a 200 kHz intrument would record reflections from the silt alone and indicate a difference in depth of a metre or more. A geophysical trace is shown in Plate 7.

4.3.3   The Side-scan Sonar Record

The side-scan sonar record is essentially qualitative, though measurements may be taken from the trace to a low standard of accuracy. In its most basic form it is an echo-sounder with its transducer tilted obliquely, and measurements made by the echo-sounder as depths

* A suppression 'decay' control is often fitted which attenuates the echo on the trace and produces a cleaner record. It must be used with caution, for weak or shallow echoes may be eliminated.

become slant ranges. The name 'oblique sonar' is sometimes used, for this reason.

The beam is invariably shaped, giving beamwidths in the order of 40° in the vertical plane orthogonal to the vessel's heading and from 1° to 2° in the horizontal plane. The transmissions may be made from a hull-mounted transducer to one side of the vessel's track ('single side-scan') or from two transducers mounted in a towed body ('fish') to both sides of the track ('dual side-scan'). These configurations are illustrated in Fig. 22. It should be noted that, although the main beam is roughly as shown, the seabed beneath the vessel is usually recorded by echoes from the side lobes, giving complete lateral coverage across the ship's track.

It will be clear from Fig. 22 that the extent of the lateral coverage of the seabed by the sonar beam depends on:

(a) the beamwidth in the vertical plane;
(b) the tilt of the beam axis to the horizontal;
(c) the height of the transducer above the seabed;
(d) the power of the acoustic transmission and of the echo which is reflected at the maximum range fringe, and the sensitivity of the receiver.

Thus, the towed-fish configuration can be more flexible than the hull-mounted, since the 'flying height' above the seabed may be selected for optimum results. Further, the fish is less subject to yawing than the fixed transducer and the depth of operation can be greatly increased without loss of resolution. A problem remains, however, of ascertaining the position of the fish relative to the 'fix' position in the towing vessel.

The side-scan sonar components are similar to those of the echo-sounder except for the transducer shape. The transducer, incidentally, performs the dual role of transmission and reception. As with the echo-sounder, the stylus continues to mark over the whole width of the trace as long as sufficiently strong echoes are being received. In the case of the side-scan sonar's inclined beam axis, echoes are received not only from the first part of the transmission to be returned (usually from the seabed) but also from the remaining, later reflections (see Fig. 23).

It will be apparent from Fig. 23 that much of the incident power of the pulse at the seabed is lost to the receiving transducer on reflection. Loss of signal strength from the outer fringes of the beam is obviated

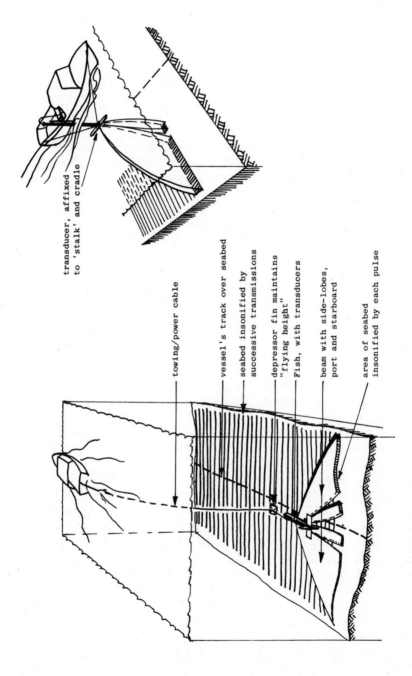

transducer, affixed
to 'stalk' and cradle

towing/power cable

vessel's track over seabed

seabed insonified by
successive transmissions

depressor fin maintains
"flying height"

Fish, with transducers

beam with side-lobes,
port and starboard

area of seabed
insonified by each pulse

Fig. 22. The usual side-scan sonar configurations: (left) the towed-fish dual mode and (right) the hull-mounted single mode

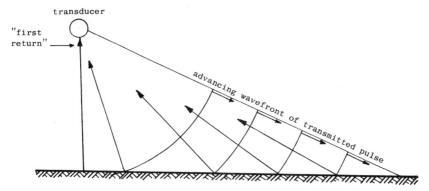

Fig. 23. A 'time' diagram of the acoustic beam. A single wavefront is shown in its
successive positions at regular intervals after transmission, together with
ray paths indicating the reflected echo return paths. The arrowheads indicate
the delay of each with respect to time after the 'first return' echo

by a varied gain control. This may be manually adjusted, or automatically increased as the return echo strength decreases across the
stylus sweep. Echoes are returned only from particles on the seabed
with surfaces normal to the incident rays. The single stylus line therefore consists of dark portions representing strong returns, decreasing
in intensity as more power is reflected away from the transducer, and
blanks representing no returns at all (i.e. acoustic 'shadows'). As the
vessel moves ahead, and successive sweeps of the stylus mark the
trace, a composite of these graded lines forms an oblique portrayal of
the seabed similar to the negative of an oblique aerial photograph.
Resolution in range is poor, an oblique range-scale, or range lines on
the trace, being provided. Resolution in azimuth (i.e. direction orthogonal to the ship's head) is unreliable due to the yawing motion of the
transducer. However, the detection capability is good with most of
the available models. Well defined 'square envelope' gated pulses are
normally used with short duration and high output levels. The com-
bination of short pulses, very narrow beamwidth and the high pulse
repetition rate made possible by the relatively short ranges (usually
50–300 m) results in a high detection probability for objects on the
seabed as small as 1 m in dimension. Sand, gravel and stones can be
identified, as can rock outcrops and wrecks.

Typical side-scan sonar records are shown in Plates 7, 8 and 9.
The features of a dual side-scan record are shown in Fig. 24.

Obviously much depends on the movement of the transducer for
the quality of the seabed coverage obtained. The trace will always

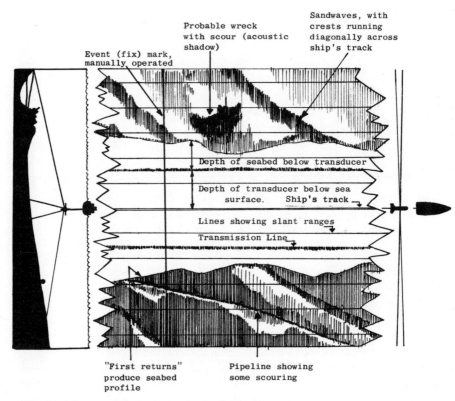

Probable wreck
with scour (acoustic
shadow)

Sandwaves, with
crests running
diagonally across
ship's track

Event (fix) mark,
manually operated

Depth of seabed below transducer

Depth of transducer below sea
surface.   Ship's track

Lines showing slant ranges

Transmission Line

"First returns"
produce seabed
profile

Pipeline showing
some scouring

Fig. 24. The common features of a dual side-scan sonar trace

indicate a smoothly progressing coverage by successive stylus sweeps
and the ship's track will always appear straight. The surveyor will
realise that this is far from the truth in practice after a few minutes'
observation of a vessel being steered along a well defined track.

The record may be made partially true to scale if the vessel speed is
adjusted to equate in scale with the trace drive speed. This is almost
impossible to achieve, however, and seldom necessary for qualitative
results. In any case, the horizontal range information is distorted
across the trace width. When scale *is* important, the echoes are best
recorded on magnetic tape, to be played back through a precision
recorder at the appropriate speed and with correction applied to the
slant ranges, or printed out by a suitably programmed computer. This
expedient is also useful for inspection of the record played back at
slightly different frequencies, much detail being revealed by the side-

bands of the centre frequency used and normally lost in the real-time record. The applications of the side-scan sonar are many and varied. For the hydrographer, it is invaluable as a means of 'seeing' between the sounding lines, enabling line spacing to be widened on occasion with confidence. The nature of the seabed material, the orientation of wrecks and the presence of sand waves and rock outcrops are all revealed and greatly assist the progress of a survey. Other applications are geological and geophysical, as an aid to the identification of sub-seabed lithology and, for the civil engineer, the inspection of proposed pipeline routes, rig sites, jetty locations, and so on.

4.3.4 The Sector-scan Sonar Record

This class of instruments, though currently under development (as, indeed, it has been for many years), illustrates the latest trends in sonar design for civil use. One prototype has been in operation since 1960 in research vessels of the Ministry of Agriculture, Fisheries and Food. This is the Admiralty Research Laboratory Sector-scan Sonar.

The fundamental design requirement is simple: complete flexibility and very high resolution in range and bearing. This can be interpreted in a variety of ways precisely as one might query the detergent manufacturers' claim that it washes whiter (than what?). To the hydrographer, the ultimate might be a capability of measurement and inspection in all directions, horizontally and vertically, between surface and seabed with a detection capability of 0·3 m targets at maximum range. The need for a sweep forward of the beam is a clear requirement for any vessel operating in uncharted waters. The 'surface-to-seabed' requirement could imply an operating range of 200 m for continental shelves and greater in other sea areas. Since supertankers are currently operating at times with an underkeel clearance of less than 1 metre (with overall draughts approaching 30 m), obstructions larger than 0·3 m are a potential hazard.

The design problems introduced by this requirement are tremendous, and a brief discussion will serve to highlight the propagation and instrumental factors involved.

To detect a 0·3 m target at 200 m requires an angular resolution of 0·1 degree. A stabilised transducer is essential, of course, but this precision also necessitates a transducer with its radiating surface having a linear dimension of 500 wavelengths. At the 'usual' frequencies of 30 kHz this is 25 m! Obviously a higher frequency must be used and the attenuation range is then reduced considerably. Transducers have some measure of self-focussing. The resolution is about $L/2$ at a range equal to $0.4L^2/\lambda$ (see Fig. 25) but this shows that the highest resolution is restricted by the dimension of the transducer (L). In fact, the use of focussed rays is essential. The resolution requirement defines the maximum width of beam as 0.3 m. The wavelength for a 0.3 m transducer of 500 wavelengths is then $0.3/500$ m and the frequency is 2.5 MHz—a totally impracticable sonar frequency. The ARL sonar has a transducer $150\lambda \times 5\lambda$, or 750 mm × 25 mm, in size, and gives a resolution of $0.3°$ (0.9 m) at a range of about 200 m with an operating frequency of 300 kHz (see Fig. 25).

The focussed beam may be produced by a paraboloid transducer-shape as implied by Fig. 25. Alternatively, the transducer may consist of a number of elements to form an *array*, the echo-data from each element being processed separately. Echoes from a target within the receiving beam of each element will arrive with phase dependent on

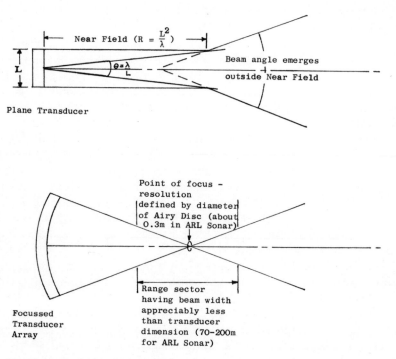

Fig. 25. The near-field characteristics of a conventional 'unfocussed' beam (top) and a focussed beam (bottom)

the direction of the target relative to the beam axis. By electronic phase-shifting the beam axis may be rotated within limits without physical rotation of the transducer array. Further refinement of resolution in the near field may be obtained by limiting the array aperture.

In the ARL sonar, a $30° \times 5°$ sector is insonified by the transmitted pulse. The receiving beam of $\frac{1}{3}° \times 10°$ scans this sector repeatedly, as described above, at a rate of 10 kHz, and the insonified sector is thus interrogated within 0·1 ms—the duration of the transmitted pulse. For this reason the term 'within-pulse scanning' is sometimes used.

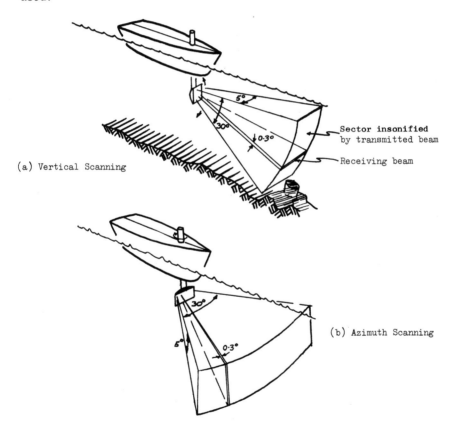

Fig. 26. The modes of operation of the ARL sonar. The depth and azimuth scanning shown above produce, in effect, a greatly improved type of side-scan capability, particularly since the transducer (and beam) in the depth-scanning mode may be rotated through $\pm 270°$ relative to the ship's head

Range resolution is theoretically 75 mm. At ranges greater than 200 m the linear resolution is lower, and typical detection ranges obtained are: small vessel, 350 m; shoal of fish, 230 m; single fish, 180 m; seabed topography out to 250 m.

Two modes are normally used: (a) with the transducer axis vertical and the insonified sector in the vertical plane to give vertical scanning, and (b) with the transducer axis horizontal and the insonified sector in the horizontal plane to give azimuth scanning. (See Fig. 26.)

The hydrographer's ideal might be a modification of the depth-scanning mode whereby the insonified sector could be directed vertically downwards, at right-angles to the ship's track, measurements derived from the receiving beam being used to provide a sounded 'lane' along the track (see Fig. 27). The width of the lane would be related to the depth, however, and an alternative solution could be found in the vertical scanning mode 'looking out' normal to the ship's head, thereby producing a consistent 200 m sweep. This mode is possible with the prototype model. The stabilised mounting has an accuracy of $\pm\frac{1}{4}°$ in pitch, roll, azimuth and tilt and $\pm\frac{3}{4}°$ in yaw stability. For these, and for the training and horizontal-to-vertical changeover mechanism, a 6 m-high well within the hull is required.

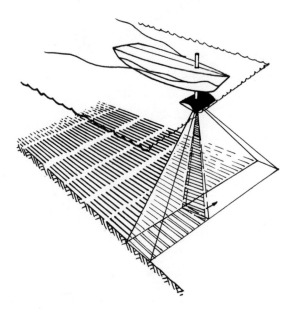

Fig. 27. The depth-scanning mode for lane-sounding

But while the engineering problems are considerable, the data-processing requirements to enable measurements of target range and size, and of seabed depths across the ship's track, are awesome.

The data-processing and recording problems are, in fact, largely unsolved. The processed data is at present displayed on a B-scan CRT arrangement, and for a permanent record a 16 mm film is taken, using a frame for each transmission, which may be played back at any speed at will. Alternatively, the returns may be recorded on tape. The conventional paper trace is impracticable since a completely new record is obtained every $\frac{1}{4}$ second, equivalent to some 2500 stylus sweeps!

The B-scan is a rectangular display showing range along one axis and bearing along another (see Fig. 28).

The filmed CRT display is excellent for surveillance and investigation purposes, but it is not satisfactory for obtaining measurements. While a real-time recording of the returns is made by tape, a digitised record in three dimensions for computer processing and automated plotting is clearly essential if the sector-scan sonar is to become a quantitative surveying instrument. It is in this respect that much development work is still required. Further, for accurate measurements and high precision, the effects of ship movement (roll, pitch, etc.) remaining after stabilisation must be corrected for and the relative bearings converted to true bearings with a very accurate gyro reference.

4.4 Underwater Acoustic Beacons

4.4.1 Other uses are made by the surveyor of acoustic waves in seawater. Chief among these are:

(i) the transmission of information from midwater or seabed instrumentation for reception by hydrophones in the surface ship or buoy (and possibly onward transmission by radio), and
(ii) the use of beacons as seabed markers for positioning, as references for checking the lane count of EPF systems and as a means of determining the depth of midwater instrument packages.

The first is essentially a matter of transmitting coded acoustic signals from transducers in the beacon; the second involves the measurement of range, bearing or both from the beacon to the receiving hydrophones.

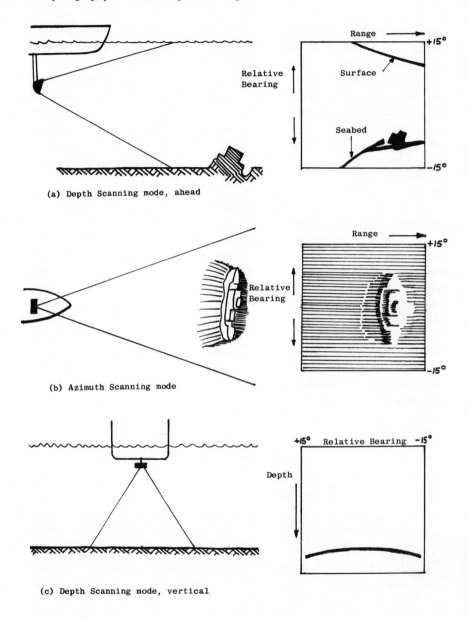

(a) Depth Scanning mode, ahead

(b) Azimuth Scanning mode

(c) Depth Scanning mode, vertical

Fig. 28. The distortion produced by the B-scan display in the various modes of operation, showing (left) the actual situation and (right) as displayed on the CRT. Note that the vessel in (a) and (b) could manoeuvre around the wreck, training the beam towards the wreck throughout

It should also be mentioned that acoustic instruments are much used by divers (as 'portable sonars'), for the data acquisition and positioning functions of submersibles, and in the hydrocarbons industry for tracing pipeline defects and blockages. These matters are not discussed here.

4.4.2   Types of Beacon

Beacons may be passive, active, or either, on command. The most simple type of active beacon (often referred to as a *pinger*) transmits short, uncoded pulses at a rate of about 1 Hz, the battery providing power for about one year. The cylindrical or spherical pressure housing may be between 0·3 and 1 m in dimension and is moored close above the seabed, preferably in a relatively flat area where the signals are not blanketed by surrounding topography. A stray line from the mooring to another sinker makes for ease of recovery by grappling, though this class of beacon is often treated as dispensable, especially in deeper waters. Alternatively, the beacon may be attached to the lowering line of an oceanographical 'cast'. (See Fig. 29.)

The beacon transmissions may be recorded on the surface vessel's echo-sounder recorder provided the frequencies are compatible. The 'echo' from the beacon will be received after reflection from the seabed and will be displayed as shown in Fig. 29.

The moored beacon may be fixed by the surface vessel in relation to EPF lane readings by making three passes at right-angles and noting the point of closest approach.

The beacon on the instrument package may be fitted with an associated reed switch so that the instrumentation is activated at a required height above the seabed, or a facility which is activated by messenger whereby the pulse repetition rate is changed to mark the end of a cast.

Beacons for the transmission of data by coded pulses are normally an integral component of the instrument package and the signals are received and processed by special associated instrumentation in the surface vessel.

Transponder beacons are the most versatile type. When placed as moored beacons on the seabed they are designed to adopt a passive mode until interrogated by a coded pulse from the surface vessel. The 'code' may be an actual characteristic pulse-train or a specific frequency of signal for which the transponder's hydrophone is 'listening'. On receipt of the interrogation pulse, the transponder

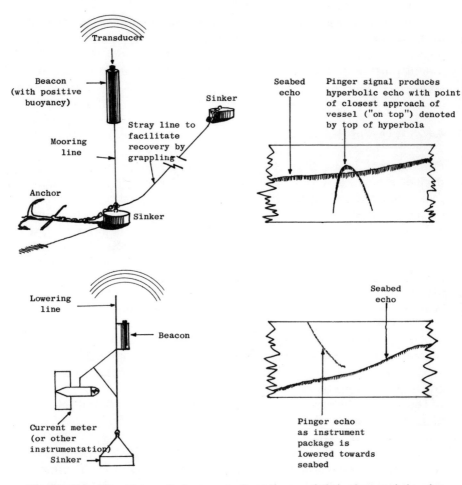

Fig. 29. Examples of acoustic beacon configurations and their characteristic echo-
sounder records

replies with its own transmission, then becoming passive until the next interrogation. Its transmission frequency is usually different from its interrogation frequency. The more sophisticated transponders are fitted with an explosive release device, activated by a special command pulse from the surface vessel. On receipt of this command, the beacon mooring is released and the beacon is free to rise to the surface, under its own positive buoyancy, for recovery. A transponder beacon is illustrated in Plate 10.

A network of three transponders may be laid in a 'triad' configuration on the seabed, fixed by successive passes of the surface vessel which uses an EPF system or (more probably) satellite or Omega fixes to ascertain its own position at the points of closest approach. When the network has been fixed, it may be used as a method of position-fixing with higher relative accuracy than the available surface system, and this method therefore has merit for large scale investigation in areas remote from the MF and microwave position-fixing facilities. Interrogation pulses are transmitted in turn to the beacons, each using an individual frequency to avoid confusion, and by timing the interval between interrogation and transponder response the position of the surface ship is fixed by two or three ranges. Owing to the appreciable time intervals involved (20 s at 15 km ranges) the speed and direction of the ship during a fix must be allowed for.

The triad transponder position-fixing method described has many drawbacks. The difficulty of fixing the beacons initially and the necessary adjustment of the 'cocked hat' produced by the time-delayed ranges are instances. Of much greater potential is the use of transponders for the dynamic positioning of a stationary vessel.

4.4.3 Dynamic Positioning

A single transponder at the seabed is sufficient for this method, but the ship-borne components are complex and sophisticated. The method is designed to enable a vessel to remain precisely on top of a particular seabed location, or to survey within a small radius of that position. Obviously these applications are suited most to the deep-sea where anchoring is difficult or impracticable; an example is the positioning of a drilling vessel (or rig) over the borehole. (See Fig. 30.)

If the ship is positioned directly above the transponder, and two hydrophones A and B are fitted so as to provide the maximum base-line length between them, the ray paths from transponder to A and B will be equal. The transponder signals will therefore arrive at both hydrophones with equal phase. If, now, the ship moves away from this position, the phase of the signals at A and B will be different, due to the unequal path-lengths, by an amount depending on the ship's distance from the 'on-top' position.

A vessel using this method will probably be fitted with bow-thruster propellers in addition to its normal propulsion screw, and the phase difference signal will be used to activate the propulsion

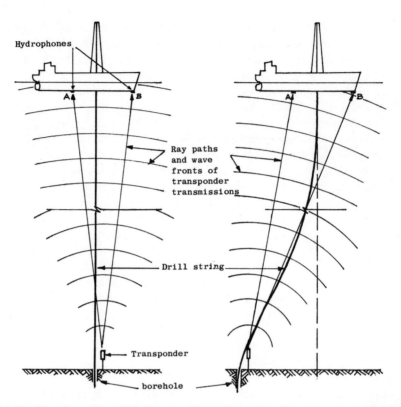

Fig. 30. The dynamic positioning configuration, with ship 'on top' and signals arriving in phase at both hydrophones (left) and with ship off station (right)

units as necessary to drive the vessel back to the null position. (It should be noted that the phase difference will be 180° when A–T differs from B–T by half of one wavelength, and the signals will arrive in phase when the ray paths differ by one or more whole wavelengths. This phase characteristic is most important in sonar design and operation. It is fundamental to the relationship between transducer size and beam directivity and in practice the amplitude of a sonar echo may be reduced or cancelled by other signals of different phase.)

The principles of dynamic positioning may be applied to the position-fixing of a surface vessel underway (see Fig. 31).

It may be seen from Fig. 31 that, for the vessel to determine its position relative to the transponder, only the length of the baseline BC and the bearing of the transponder need to be known. The length BC is given by AC sin θ, and the length of AC may be obtained by

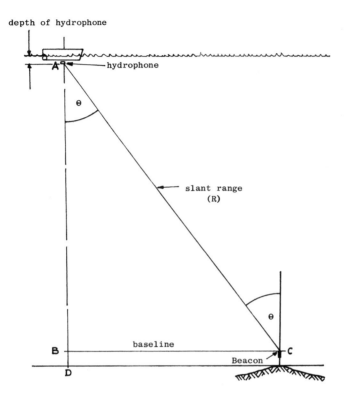

depth of hydrophone

hydrophone

θ

slant range
(R)

θ

baseline

B

D

Beacon

C

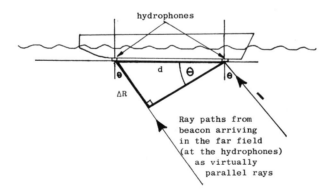

hydrophones

d

θ

θ

θ

ΔR

Ray paths from
beacon arriving
in the far field
(at the hydrophones)
as virtually
parallel rays

Fig. 31. The geometry of single-beacon position-fixing in the fore-and-aft plane only

timing the interval between the transmission from the transponder and reception at the hydrophone. However, this can only be determined if the time of interrogation is also known, together with the transponder's internal delay before transmitting (as in the case of triad systems). In fact, a simple beacon may be substituted for the transponder and the need for interrogation is averted if *differences* of range (ΔR) can be measured, rather than total range.

The range difference can be computed if the sound velocity (c) is known and if the difference in arrival times (Δt) of a wave front at each of two hydrophones is measured. Δt can readily be measured in terms of the difference in phase at the two hydrophones. Then

$$\Delta R = c\Delta t \quad \text{and} \quad \sin \theta = \frac{c\Delta t}{d}$$

where d is the distance between the hydrophones.

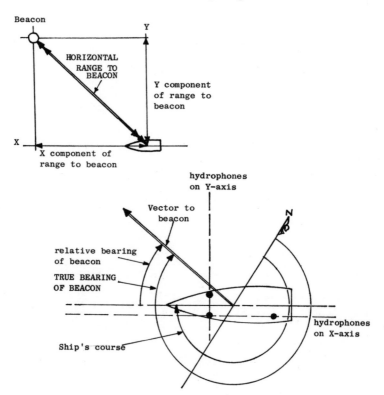

Fig. 32. The derivation of range and bearing from the beacon

Since R is no longer measured, knowledge of the depth AB is required and this may be obtained from the echo-sounder, allowing for the length of mooring line BD. Then

$$BC = AB \tan \theta$$

and, since θ is small,

$$\tan \theta = \sin \theta \quad \text{and} \quad BC = \frac{ABc\Delta t}{d}$$

If a third hydrophone is fitted, making an angle of 90° with the first two, a similar computation may be carried out for the baseline of the other axis, thereby avoiding the measurement of bearing to the beacon (see Fig. 32). The X-axis is, of course, along the heading of the vessel and a gyro reference is required to convert from relative to true directions. Further, a vertical reference is required to eliminate errors due to pitch and roll which upset the ΔR triangle.

These systems are accurate to about $0.01 \times$ water depth within a radius of $0.2 \times$ water depth from the beacon, and the position data is presented on a PPI (plan-position indicator) cathode ray tube display, or converted to digital read-out for printing or plotting.

This principle is of value in deep water surveys using a towed-fish for echo-sounder, side-scan or geophysical operation. The fish enables high resolution records to be obtained, free from surface noise made by the ship and waves, and from the yawing, pitching and rolling motion experienced at the surface. The great problem is in determining the position of the fish relative to the ship, and may be solved by attaching a beacon to the fish, together with an echo-sounder to give the altitude of the fish above the seabed. The system can be simplified by assuming that the fish is directly astern of the vessel, therefore requiring hydrophones in the X-axis only.

CHAPTER FIVE

Miscellaneous Hydrographic Operations

5.1 General

The previous chapters have described how the surveyor charts the topography of the sea environment. The use of sonar to measure depths and scan the seabed for the smallest features has been discussed. In this chapter the remaining field operations are described in the context of hydrography as interpreted in this book—that is, distinct from oceanography and geophysics and geology. However, it is difficult to isolate hydrography from these entirely. If the hydrographic purpose is seen to be to acquaint the user with information sufficient for his further work in the charted area, the survey necessarily includes observations which border on the oceanographic and geological. An example covered in the last chapter may be found in the side-scan (or sector-scan) sonar, whereby the surveyor finds the information invaluable for his chart of the seabed topography, and the geologist, using the same record, obtains useful clues from the seabed geomorphology revealed.

In this chapter, therefore, the geological sampling of seabed material is described as an aid to the mariner and engineer using the chart, and the measurement of the horizontal and vertical movement of the seawater mass for the same reason. The sweeping by wire of rocks, wrecks and other isolated features, also described, is a purely hydrographic operation which complements the bathymetry.

It is not intended to describe rudimentary techniques such as sounding by lead line or depth-pole, since these are largely self-explanatory and little used in industrial hydrographic surveys.

The surveyor might well be required to make observations other than those described here. Examples are the tracking of dye or radio-active tracers for pollution or sediment transport studies, plankton sampling for biologists and meteorological and wave observations for civil engineering projects. In these cases he will be the position-fixing expert and data acquisition man. The data acquired will usually be assessed or processed by scientific personnel and the surveyor will be working under their specific instructions. The surveyor should appreciate his role in these operations as a member of the team. This is particularly relevant in geophysical surveys which are essentially extensions of the hydrographic survey and in which hydrographers are increasingly involved.

5.2 Water Flow and Solids in Suspension

5.2.1 Tides are fundamental to the surveyor and engineer. Their importance in the reduction of soundings to a datum level has already been mentioned (para 2.4). It is assumed that the reader is familiar with the astronomical forces which generate the tidal wave, and with the cyclic nature of the rise and fall of the tide and the Spring and Neap phases of the tidal régime.

For the water level to rise and fall there must be a flow of water in a horizontal direction attendant upon the vertical movement. As the determination of the datum level and predictions of tidal heights for the future take account only of the astronomical and known topo-graphical (i.e. shallow water) influences, so the associated *tidal stream* is defined by those same factors. In other words, the horizontal flow caused by the tide-generating forces will be greatest at half-tide when the rate of change of level is greatest, and least at high or low water when there is no change of level. The fact that some horizontal move-ment may be observed at high and low water, and that the ebb and flood tides experience different rates, indicates the presence of a *current* which may be due to river flow, topographical features which trap, divert or restrict the water movement, or to a difference of atmos-pheric pressure or water density over an area creating a sloping sea surface and consequent gravity or gradient currents.

Observations of tidal range and tidal streams and currents are clearly a prime requirement in the design of underwater pipelines, offshore platforms and harbour works. The ability of the water movement to carry with it particles of seabed material is of equal interest. While observations of salinity and temperature will enable the oceanographer to estimate this effect, the direct observation of the solid matter in suspension in the seawater is of positive value, particularly when assessed in conjunction with water flow and seabed sampling observations.

5.2.2 Observations for Average Water Flow

There are two major methods of measuring the water flow, and each is suitable for one of the prime user requirements.

The mariner's interest is in the average total flow experienced between the surface and a specified depth, as by a vessel of a certain draught. For this, the *pole logships* may be used, or a series of *current meter* or *float tracking* observations may be made at intervals of depth through the water column. The latter method, providing information on the flow direction and rate at predetermined depths, is better suited to the engineering requirement for individual observations, since to obtain the average flow necessitates integration and averaging of the results, whereas the pole method obtains the average flow direct.

The nature of tidal streams and their observation by the pole logships are fully and well described in the *Admiralty Manual of Hydrographic Surveying*, Vol. II. Suffice it here to say that the method consists of the streaming of a vertically floating pole from the control vessel at intervals of, say, 30 min over one or more tidal cycles and, from the distance and direction travelled by the pole in a timed interval (one or two minutes), determining the rate and direction of the flow.

For the *average* flow to be measured, the pole must be weighted so that it floats to the depth over which the measurements are required and the amount allowed to protrude above the surface (i.e. *freeboard*) kept to the minimum consistent with visual tracking of the pole, to minimise the influence of wind on its movement. The pole may be tethered by distance-line to the stern of an anchored vessel, in which case the pole is fixed at the beginning and end of each run by bearing and distance relative to the stern which, in turn, must also be fixed. Alternatively, and better for observations from a small launch, the pole may be free-drifting and tracked by closing the pole at intervals and fixing when the boat is alongside. In either case, arrangements

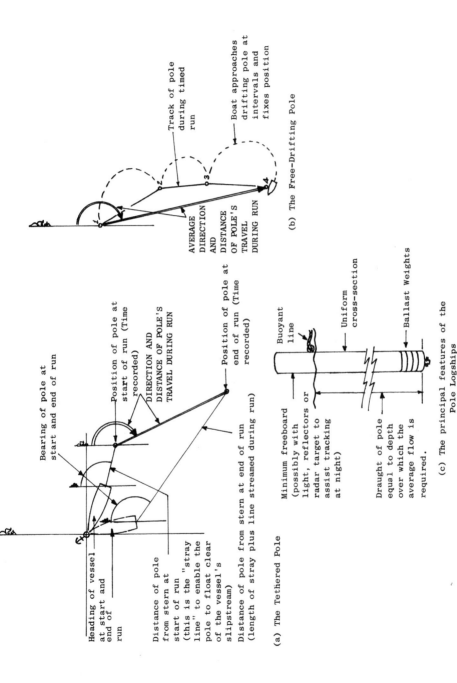

Track of pole during timed run

Boat approaches drifting pole at intervals and fixes position

AVERAGE DIRECTION AND DISTANCE OF POLE'S TRAVEL DURING RUN

(b) The Free-Drifting Pole

Bearing of pole at start and end of run

Position of pole at start of run (Time recorded)

DIRECTION AND DISTANCE OF POLE'S TRAVEL DURING RUN

Position of pole at end of run (Time recorded)

Heading of vessel at start and end of run

Distance of pole from stern at start of run (this is the "stray line" to enable the pole to float clear of the vessel's slipstream)

Distance of pole from stern at end of run (length of stray plus line streamed during run)

(a) The Tethered Pole

Buoyant line

Minimum freeboard (possibly with light, reflectors or radar target to assist tracking at night)

Uniform cross-section

Ballast Weights

Draught of pole equal to depth over which the average flow is required.

(c) The principal features of the Pole Logships

Fig. 33. The pole logships methods of water flow measurement

must be made to obtain the track of the pole by night as well as by day since the sequence of observations must not be broken during a tidal cycle. Wave and wind action on the pole may be stronger than a weak water flow, and it is therefore important to observe in calm conditions.

This method (and float-tracking described in para 5.2.3) is expensive, since a vessel and entire crew are occupied for a period of at least 13 hours. This single tidal cycle will produce only the most crude results. As in the case of tidal height observations for the purposes of analysis and prediction, 19 years is the ideal period, a lunar month produces tolerable results, and a single cycle can only give the most basic indication. However, as an aid to navigation this 'indication' is sufficient, and a simple but effective analysis described in the *Admiralty Manual* enables adequate predictions to be made. For scientific research, a statistical harmonic analysis would be required, justified only by extensive observations, but this would be likely only for the observations of water flow at specific depths as described below.

5.2.3 Observations of Water Flow vs. Depth

In shallow waters, improvised floats of many kinds may be tracked to observe water flow. A common type consists of a plywood vane with sinker, suspended beneath a small pellet-type float. In deeper waters, parachutes or drogues may replace the vane, or special, neutrally buoyant floats ('Swallow' floats) may be tracked by acoustic means as they drift at their natural depths. These devices are illustrated in Fig. 34.

The recording current meter has the great advantage over all these of being self-contained, leaving the survey vessel free to progress other work. Basically, almost all current meters employ similar devices for the measurement of water flow, only the means by which the data is recorded being of differing design. An impeller or rotor is attached to the meter body. A fin on the body maintains the water flow along the axis of the instrument and the impeller facing the flow. The impeller is so designed that its rate of rotation is proportional to the rate of water flow. Direction is observed by a magnetic compass within the meter body.

These points are illustrated in Fig. 35, and an example of a Savonius rotor recording current meter is shown in Plate 11.

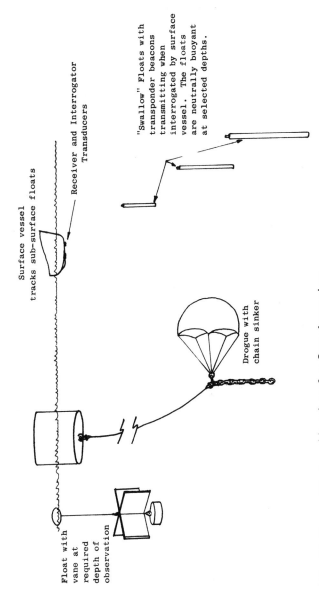

Surface vessel
tracks sub-surface floats

Receiver and Interrogator
Transducers

"Swallow" Floats with
transponder beacons
transmitting when
interrogated by surface
vessel. The floats
are neutrally buoyant
at selected depths.

Drogue with
chain sinker

Float with
vane at
required
depth of
observation

Fig. 34. Types of float used in sub-surface flow observations

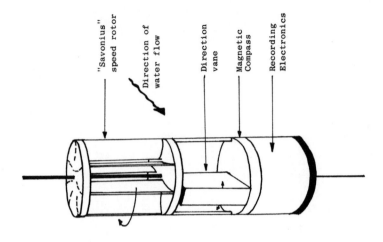

"Savonius" speed rotor

Direction of water flow

Direction vane

Magnetic Compass

Recording Electronics

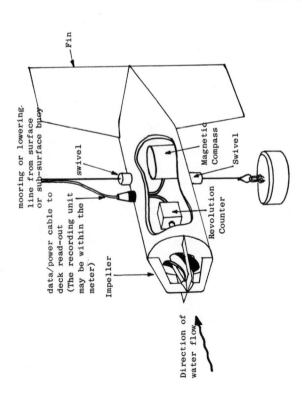

Fin

mooring or lowering
line from surface
or sub-surface buoy

data/power cable to
deck read-out
(The recording unit
may be within the
meter)

swivel

Impeller

Magnetic Compass

Swivel

Revolution Counter

Direction of
water flow

Fig. 35. The principal features of current meters

Referring to Fig. 35, several important points arise:

(i) The rotor must be protected from debris in the water.

(ii) The instrument body must be watertight. The rotor is often magnetically coupled to the revolution counter to avoid penetration of the body.

(iii) The body must be freely suspended to allow it to take up the direction of the water flow.

(iv) If the current meter is direct-reading, a means must be provided of passing the rate and direction data to a surface read-out unit. If the meter is of the recording type, a recording device must be incorporated.

The direct-reading meter cannot be left to operate unattended. The rotation of the impeller is usually designed to generate a voltage which is carried by cable (often serving also as the lowering line) to a dial or digital read-out on deck.

The direction of the instrument axis relative to compass north may be measured in a number of ways. A common method is to connect a potentiometer into a bridge circuit. The resistance of the potentiometer is determined by the compass needle which acts as the setting-arm, and the potentiometer resistance is nulled by the adjustment of a second resistance in the bridge in the deck unit. In the recording current meter the impeller rotation and the variations in position of the compass needle are caused to generate voltages or coded pulses or changes of a resistance-ratio which are recorded within the instrument. There are many types of recording device. Some models employ magnetic tape on which binary signals are recorded for later processing. Others incorporate a graphic recorder with the pen positioned over the chart by the compass and lowered to the chart paper by revolutions of the impeller. The record then consists of a number of dots, their position on the chart signifying direction and their frequency of occurrence being proportional to the rate. The chart paper must be graduated in time, of course, and driven by an accurate clock.

Recording current meters can operate unattended for periods of several months, depending on the capacity of the recording medium and battery life. One month is usually the maximum time required for one set of observations.

The meter is capable only of observing the water flow at one depth, and various configurations are used for obtaining data at several depths concurrently. (See Fig. 36.)

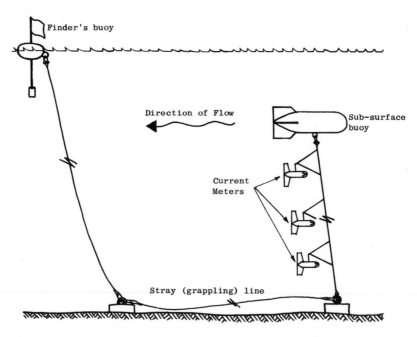

Fig. 36. A typical current meter array

5.2.4   Observations of Solids in Suspension

The quantity of solid matter (sediment) which can be held in suspension in seawater can be usefully interpreted, together with water flow data and seabed samples, to assess the seabed stability or mobility. Applications are in feasibility studies of all kinds, for example hydraulic models, sewerage schemes, dredging and reclamation works, jetty and breakwater construction and other engineering projects. Surveys to determine the consequential effects following such projects will also include observations of the suspended solids, for example in the surveillance of scouring around the piling of an offshore platform or beneath a pipeline.

Material of any size may be transported by the water flow provided the rate of flow is sufficient. Ocean waves breaking on an exposed coast are capable of lifting and moving great boulders, and even the most gentle stream or current will have a quantity of sediment in suspension close above a sandy or silty seabed. Similarly, material will be deposited whenever the rate of flow falls to a level which is unable to support it.

Whilst observations of salinity (i.e. dissolved salts) over a period of time will provide a comparative indication of the suspended matter (both suspended and dissolved matter affecting the electrical conductivity which is widely measured for the determination of salinity), the solid matter is best assessed by the turbidity or transmissivity of visible light.

The transmissivity instruments measure, by photo-electric cell, the amount of light received over a fixed path-length from a built-in light source. The backscatter type subjects the sample volume of water to pulses of light, the intensity of that light reflected by the particles being measured by a photocell in the receiver (see Fig. 37).

It should be noted that the concentration of pollutants or phytoplankton may be determined and their nature identified by inserting filters in the turbidity meter and noting their spectral range. Two receivers may be used, one with and one without filters. The difference in read-out between the two receivers is then an indication of the concentration of those substances having the spectral range isolated by the filters. The injection of fluorescent substances into the water sample enables the relative intensities of fluorescent irradiation to be measured for more positive analysis and interpretation.

5.3 Seabed Sampling

A great variety of seabed samplers are available to the surveyor and are documented in many oceanographical texts.

The requirement is invariably for the acquisition of *undisturbed* samples and the designer attempts to come as close as possible to this impossible criterion within the limits of penetration desired and the type of seabed involved. The degree of disturbance which is acceptable will vary with the purpose for which the samples are intended. An engineer may consider it important that the water content of the seabed material be determined so that realistic tests may be made of strengths, load-bearing properties and stability. A biologist will be interested in preserving the organic content, while the geologist will usually be satisfied with a sample which reveals the constituent materials of the seabed, though he will wish the sample to be undisturbed.

Three basic types of sampler are used (see Fig. 38).

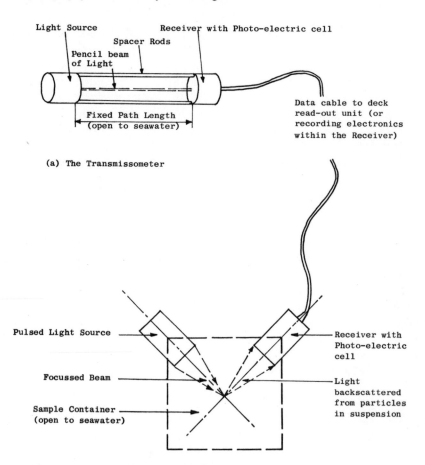

(a) The Transmissometer

(b) The Backscatter Turbidity Meter

Fig. 37. The principal methods of measuring the solid matter in suspension

The *grab sampler* obtains a reasonably undisturbed sample in silt, sand or loose gravel in a precise location, and certain types retain the water content of the seabed material. Penetration is to a fraction of a metre only.

The *dredge* is used, as the name implies, to scrape loose material from the surface of the seabed and, depending on size and type, will cope with fine sediment, stones or boulders.

The *corer* is dropped in a precise location and obtains a sample to a depth of from one to several tens of metres depending on the type of corer and the nature of the seabed. The simple corer comprises a

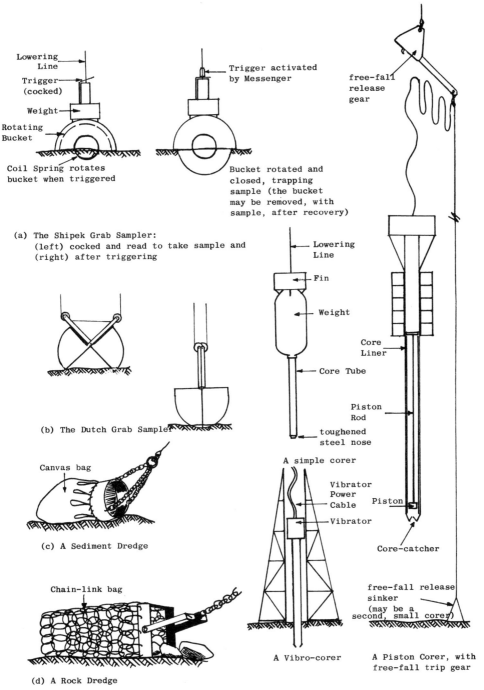

Lowering Line

Trigger (cocked)

Weight

Rotating Bucket

Coil Spring rotates bucket when triggered

Trigger activated by Messenger

Bucket rotated and closed, trapping sample (the bucket may be removed, with sample, after recovery)

(a) The Shipek Grab Sampler: (left) cocked and read to take sample and (right) after triggering

(b) The Dutch Grab Sampler

Canvas bag

(c) A Sediment Dredge

Chain-link bag

(d) A Rock Dredge

Lowering Line

Fin

Weight

Core Tube

toughened steel nose

A simple corer

Vibrator Power Cable

Vibrator

A Vibro-corer

free-fall release gear

Core Liner

Piston Rod

Piston

Core-catcher

free-fall release sinker (may be a second, small corer)

A Piston Corer, with free-fall trip gear

Fig. 38. The various types of seabed sampling device

weight with detachable core-tube maybe one metre in length. A *core-catcher* closes when the corer is lifted, to prevent the material from escaping or from being washed out. A *core-liner* is a plastic inner tube which may be drawn from the core tube on recovery and used to retain the sample undisturbed until required for laboratory tests. Corers of this type tend to cause changes in the sample through compression within the tube, and a *piston-corer* is designed to obviate this and achieve greater penetration. The corer may be dropped in shallow water, lowered to the seabed, or released by a trip mechanism and allowed to fall freely for the last few metres. More sophisticated developments include a vibrating head to the corer to achieve greater penetration, or to pierce hard material. No corers are capable of penetrating rock.

The grab, dredge and corer are probably the only devices which the surveyor might be required to operate. Deeper samples, through harder material, are obtained by boring and drilling. The samples are disturbed, of course, and the equipment is found usually only on vessels or platforms specially designed for this work.

5.4   Seawater Sampling

The properties of seawater are of interest to the oceanographer and other scientists in the service of the engineer as well as for pure research. The surveyor may be called upon to provide samples of seawater for laboratory analysis and for this purpose a simple water sampler is required which incorporates the following features:

(i)   a clean and chemically inert container which will not contaminate the sample;
(ii)   a method of closing the container at the required depth;
(iii)   the ability to draw off the sample on deck.

The 'Nansen' reversing water bottle is the most commonly used—and most refined—sampler for all depths. When triggered by the messenger, the bottle is allowed to fall 'head over heels', while at the same time the upper and lower mouths of the bottle are closed and the water temperature at the point of reversal is obtained by means of associated thermometers.

The ubiquitous 'messenger' is used extensively for the remote activation or control of underwater instrumentation. It is simply a

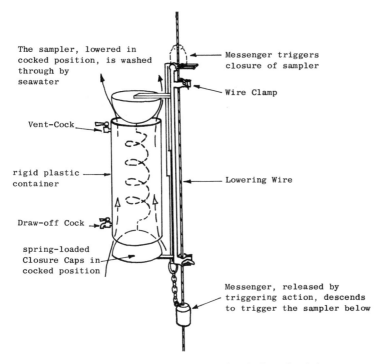

The sampler, lowered in cocked position, is washed through by seawater

Messenger triggers closure of sampler

Wire Clamp

Vent-Cock

rigid plastic container

Lowering Wire

Draw-off Cock

spring-loaded Closure Caps in cocked position

Messenger, released by triggering action, descends to trigger the sampler below

Fig. 39. A simple water sampler (suitable for shallow depths)

small weight which is slotted around the lowering wire on deck and, when it reaches the instrument, triggers the operation required. In an oceanographical cast of several instruments at intervals of depth down the wire, each instrument has a messenger which is released on arrival of the next messenger above. This goes down to activate the next instrument below, which is caused to release its messenger, and so on until all instruments have been triggered.

5.5 Wire Sweeps

When the survey area has been investigated by echo-sounder and sonar, certain features may have been located and it becomes necessary to ascertain their least depth exactly. Such features may be pinnacle rocks or wrecks and the like, and it is conceivable that the remote sensing instruments have not detected the highest points. These may be a small stanchion, a mast or a piece of jagged metal or rock, in

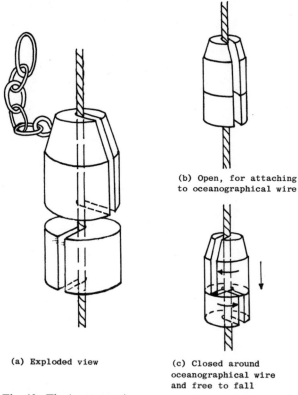

(b) Open, for attaching
to oceanographical wire

(a) Exploded view

(c) Closed around
oceanographical wire
and free to fall

Fig. 40. The 'messenger'

many cases reflecting too little of the acoustic power to survive as an echo on the sonar record. Alternatively, a channel may have been selected from the bathymetry as a safe route to a certain depth for users of the chart, and the absence of any obstructions projecting from the seabed must be assured. The wire sweep is a means of physically ascertaining the safe depth of water in either case.

The actual form of sweep is largely a matter of commonsense for the surveyor and seaman. Clearly, the following criteria must be met:

(i)   A rod, bar or line must be towed horizontally through the water at a known depth (known, that is, below the datum level: a knowledge of tidal height throughout is implied).

(ii)   The depth of the rod, etc., must be such that any obstruction is both fouled and cleared so that the depth of the obstruction is 'bracketed' within acceptable limits.

(iii) In the case of the safe channel, the entire area concerned must be covered by the sweep to the depth recommended as safe. In the case of an obstruction, a generous area around the obstruction must be swept to ensure that the highest point has, indeed, been located. (In ports and their approaches, it is often the practice to tow a sweep of heavy chain cable along the bed in an attempt to dislodge and flatten any obstructions in its path.)

(iv) The position of any obstruction swept must be fixed precisely.

In practice, the swept path is made as wide as possible for economy of effort consistent with adequate accuracy. A fine line or piano wire is commonly used, suspended from two strong, weighted depth-lines. Intermediate buoys (small floats) may be placed at intervals to avoid too great a sag between supports. A single-ship sweep may be used, the line being suspended from beneath the bow and stern, but the width of the swept path is then severely limited. Two vessels are better used, the surveyor in one fixing position and controlling the operation and the crew of the other maintaining a specified bearing and distance from the first. A typical situation is shown in Fig. 41.

A slow speed of advance is necessary, to avoid the vessels coming up all-standing if an obstruction is fouled and to avoid the sinkers and depth lines streaming astern. Some speed is necessary as distinct from drifting, to enable the vessels to maintain station and the intended line of the sweep. The vessels therefore head into the tidal stream or current, or the wind, whichever is the stronger, and the true depth of the sweep is found by pressure or acoustic transducers on the sinkers, with their read-out on deck.

The usual methods of position-fixing are used, but it is good practice to place a buoy up-tide from the obstruction as a visual reference.

The sweep is set, in the first instance, to the least depth of the obstruction by echo-sounder or sonar, with the intention of fouling on the first run. The obstruction is then positively located by manoeuvring directly over the fouled wire, using the direction of the sweep wire as a guide. On the next run, the depth of the sweep is decreased with the intention of just clearing the obstruction. Ultimately, in this manner, the clear and foul depths (reduced for tidal height) should be about 0·3 m apart—the maximum precision which may be expected from this rig. (If intermediate floats have been used, the float nearest to the obstruction will dip beneath the water on fouling and the other floats will adopt an arrowhead formation, pointing towards the position of the foul.)

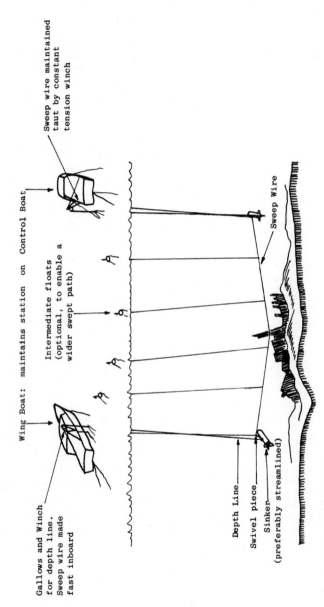

Sweep wire maintained taut by constant tension winch

Wing Boat: maintains station on Control Boat

Intermediate floats (optional, to enable a wider swept path)

Sweep Wire

Gallows and Winch for depth line. Sweep wire made fast inboard

Depth Line

Swivel piece

Sinker (preferably streamlined)

Fig. 41. A two-ship sweep

In the case of a clearance sweep, sufficient runs at the (reduced) safe depth are made to cover the area, unless, of course, some obstruction is found.

The above remarks summarise the fundamentals of sweeping by wire. Much of the detail consists of seamanship and experience, and the reader is recommended to study the appropriate chapter of the *Admiralty Manual of Hydrographic Surveying*, Vol. II.

CHAPTER SIX

The Processing and Presentation of Hydrographic Data

6.1 The Track Chart

6.1.1 Hydrographic surveys almost invariably result in a chart of the area portraying the acquired data as topographical and bathymetric information, and possibly, in addition, tabulated information on such matters as currents and tidal streams. A report should be part of the standard presentation, pointing out details relevant to the purpose of the survey and amplifying the other documents rendered as necessary.

The compilation of the chart, tables and report is made from many different sources, for example, the echo-sounder and sonar records, the seabed sampling log, the tidal stream analysis sheets, and so on.

All the acquired data is meaningless without the position in the survey area to which it relates, however, and common to all the sources of environmental data is the plot of the vessel's position throughout the survey—the track chart. Transparent plastic overlays to the track chart are used to insert the bathymetric and other data, as many overlays as are necessary being used to portray the information clearly and to separate one type of information from others where required.

The track chart is required from the beginning of the survey and is progressed until the completion of the fieldwork. The intersections of the grid in use are pricked at intervals of about 150–200 mm on the scale of the survey, and geographical intersections may be shown also—particularly if the fair sheet is to be given a graduated border. The land surveyor is familiar with the drawing up of master sheets on which the grid is constructed—an operation greatly facilitated by millimetre gridded plastic sheets obtainable from the Ministry of Defence Hydrographic Supplies Establishment at Taunton. Stable plastic transparent sheets of 0·1 mm thickness (e.g. Ozatex) are recommended for the track chart and all the associated overlays and underlays.

If visual methods of position-fixing are to be used, the track chart may need only to show the control points onshore which will be observed. (Whatever the method of positioning, it is as well to show these points as visual check-fixes will almost certainly be required from time to time.)

6.1.2 Lattices

Fixing will be greatly facilitated if lattices of constant subtended angle are prepared, either on the track chart itself or on underlay sheets. While it is true that a lattice of more than, say, three 'families' of subtended angles can be confusing and conducive to error in plotting fixes, the surveyor's choice of fix need not be limited if several different underlays are prepared. The advantages are considerable, as described in para. 3.2.

It is often said that lattice construction is only justified for areas subjected to repeated re-survey, such as dredged channels. However, lattice calculations are simple (see Chapter 7) and the plotting by beam compass presents no problems if the control points to be used appear on the master sheet. A major advantage of the fixed angle plot lies in its use for large scale surveys, where the arms of the station-pointer might not be long enough to span the distance from the fix to the control points, and where the track chart would be unnecessarily large if the control points were to be included. In such cases, the arcs must be drawn with the aid of splines or curves, and the arcs defined by numerous co-ordinated points along their length. The procedure is then quite laborious, though the use of standard circle sheets (supplied by the Hydrographic Supplies Establishment) eases the work considerably.

The remarks immediately above must be qualified in the light of the increasing application of computers and X-Y plotters. A computer-plotted arc is infinitely more precisely drawn than by hand.

Although it is usually unnecessary to consider projection scale errors on areas bounded by the limits of visual (and even microwave) fixing, it is a simple matter to program the computer to accommodate scale factors, and the resulting ellipses are even more accurate than hand-drawn arcs.

Where fixing is by two-range or hyperbolic EPF system the incorporation of projection scale factors is essential. The surveyor will normally construct his lattice in zones, allowing for the changes in scale factor in steps from zone to zone. In these cases computer-plotting is really justified. The drawing by hand of all types of lattice is explained in the *Admiralty Manual of Hydrographic Surveying*, Vol. I.

6.2   The Processing of Bathymetric Data

6.2.1   'Soundings', reduced for the height of tide above the chosen datum, are plotted as spot-depths along the track of the survey vessel.

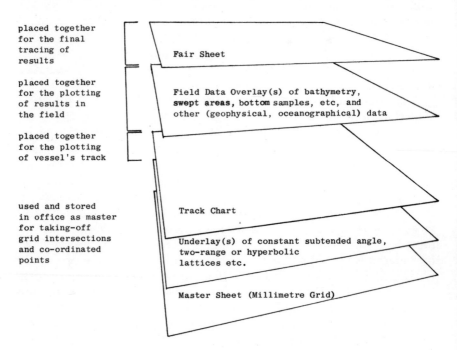

Fig. 42. The various sheets required for the hydrographic survey

As other data—e.g. magnetic, gravity and seismic—may have been acquired concurrently, a separate overlay tracing is normally used for each.

In the conventional way, the spot soundings at intervals of, say, 10 seconds (depending on the survey scale) along the track represent only about one per cent of the total information acquired and recorded on the echo-sounder trace. The spacing of the soundings is only as accurate as the eye of the draughtsman permits, and the reading-out of each sounding one by one is obviously prone to error.

If automated methods are available the (unreduced) soundings will have been recorded on punched paper or magnetic tape, and the realtime plotting of the soundings by X-Y plotter is prevented only by the absence of tidal reductions. In these circumstances, the bathymetric record is played back on completion of the day's field work, the tidal heights being injected manually or the digitised tide-gauge record being incorporated as a correction tape with the echo-sounder tape. The corrected record can then be fed to the X-Y plotter.

Even with the computer capability described above, the amount of depth information which can be plotted is limited by the space available and is no greater than in the manual processing method. It is preferable to program the plotter to print contours rather than spot depths, or contours together with spot depths indicating maxima and minima. If a very close vertical interval is adopted, the contoured chart is superior by far to the hand-drawn version. The maximum accuracy is preserved, free from human error, and the contours convey a three-dimensional effect which portrays the nature of the seabed topography quite dramatically.

6.2.2 The Conventional (hand-drawn) Method

At the fieldwork stage of a bathymetric survey, the following action is taken at each fix:

(i)   A record is kept of the time at the instant of the fix, the sequence number of the fix and the details of the fix (e.g. angles observed, EPF system lane readings, etc.).
(ii) The fix is plotted and joined by a straight line to the previous fix to represent the track followed by the vessel.
(iii) The instant of the fix is marked on the echo-sounder trace.

Tidal heights above the chosen datum are read from a nearby tide-scale and recorded against time for later correlation with the field records.

The echo-sounder trace, the sounding log and the track chart are removed to the survey chartroom on completion of the day's field-work.* The record of tide-readings is used in conjunction with the sounding log to derive the reduction at the various fix numbers. The tidal reduction is then plotted by fix number at intervals of about 0·3 m along the echo-sounder trace and the tidal curve is thus reconstructed on the trace (see Fig. 43).

The reduced soundings are then read off the trace by placing the zero of the reading-off scale against the tide curve. The soundings indicated are inked in along the track at the required spacing. When all the lines of sounding have been inked in, the surveyor in charge contours the chart, using his experience to interpret the trends and deciding upon any interlines or other verification fieldwork required to resolve ambiguities or fill in gaps in the coverage.

Finally, because of the gradual progression of the bathymetric overlay by different hands, it is virtually certain that a fair tracing will have to be made for presentation to the client. A cursory appraisal of the above remarks will indicate the scope for errors occurring at every stage, and the need for meticulous care at all stages of data processing.

The process is summarised in Fig. 44.

6.2.3 Automation

Positional, depth and other data are increasingly output by the various instruments in binary decimal form.

Digitised Positioning Data
The automation of the track-plotting operation from digital positioning data is relatively simple. Given the parameters of the spheroid and projection in use, and the co-ordinates of the shore stations of

---

* The importance must be stressed of plotting the reduced data on the same day as the field-work, and by the surveyor responsible for its acquisition. Memory plays an intangible but absolutely vital part in the correct transference of the echo-sounder data to the bathymetry overlay. Spurious echoes from weed, side-lobe returns, fish, etc., will have been noted by the surveyor at the time. He will also have noted whether further investigation was needed at that time, and will know whether the doubtful echo is to be rejected or not at the inking-in stage.

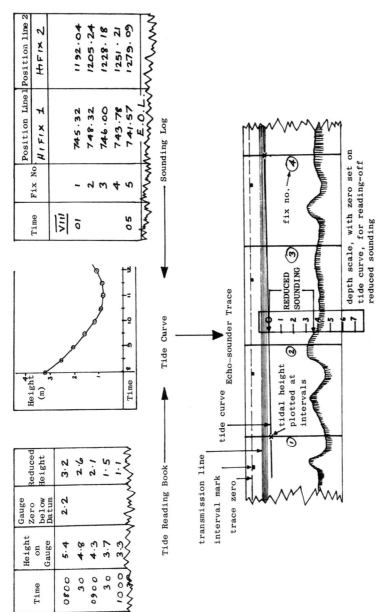

Fig. 43. The steps involved in the reduction of soundings

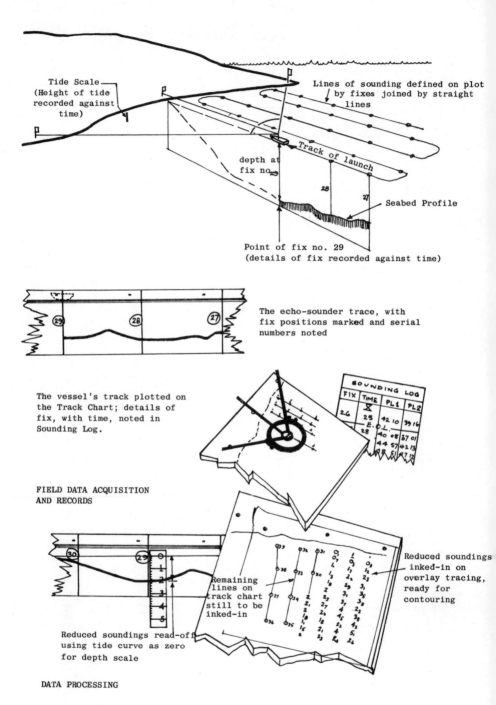

Tide Scale
(Height of tide
recorded against
time)

Lines of sounding defined on plot
by fixes joined by straight
lines

depth at
fix no.

Track of launch

Seabed Profile

Point of fix no. 29
(details of fix recorded against time)

The echo-sounder trace, with
fix positions marked and serial
numbers noted

The vessel's track plotted on
the Track Chart; details of
fix, with time, noted in
Sounding Log.

SOUNDING LOG

FIX	TIME	PL1	PL2
26	25	42 10	39 16
	E.O.L.		
28	40 08	37 01	
	44 57	42 13	
		47 17	

FIELD DATA ACQUISITION
AND RECORDS

Remaining
lines on
track chart
still to be
inked-in

Reduced soundings
inked-in on
overlay tracing,
ready for
contouring

Reduced soundings read-off
using tide curve as zero
for depth scale

DATA PROCESSING

Fig. 44. The stages of the sounding operation

an EPF system together with the propagation velocity and frequency of operation, a program may be written for the conversion of lane-counts to geographical or grid terms and the track may then be plotted direct on an X-Y plotter. Due to the variations in EPF system read-outs caused by propagation anomalies, lane-slip, voltage fluctuations and noise, the program usually caters for the sampling of the system read-out at short intervals (say, 0·1–1 s) and a smoothed print-out by the X-Y plotter at intervals of 0·5–1 min depending on the scale of the plot. The printing of consecutive fix numbers or times is easily catered for.

Digitising the Depth Record
The difficulty of digitising the essentially analogue echo-sounder data has delayed the automation of hydrographic operations relative to other branches of technology. Progress in this area is now being made in all but the smallest survey organisations, spurred on by the demand for the measurement of an ever-increasing number of parameters, with greater accuracy and in a shorter time than was required previously for nautical charting surveys.

The essentials of digital echo-sounders are an amplitude gate, a time gate and a counter. The transmission of the acoustic pulse is synchronised with the triggering of the electronic counter. The returning echo is subjected to two tests before being permitted to stop the counter. First, it must be of sufficient strength to pass over the threshold of the amplitude gate, and, second, it must arrive within the interval during which the time gate is open. The seabed echo is therefore assumed to be the strongest signal returned. The threshold of the amplitude gate is set by automatic gain control, so that weak bottom returns are not rejected, and the time gate operation is synchronised with the seabed echo of the previous sounding cycle, the open time allowing for the maximum expected variation in depth between echoes. The digital circuitry thus 'locks on' to the seabed echoes for as long as they are received.

When the echo has passed through the gates it is made to stop the counter. The count is passed to a data register and the counter is reset, ready for the next transmission.

If, for some reason such as aeration under the transducer, an echo is not received, the data register retains the last received measurement and the counter is reset so as to be ready to accept the echo of the next sounding cycle. If after a predetermined number of cycles the echo is still not retrieved, the read-out register is returned to zero, thus indicating failure, and the counter is reset, awaiting rectification of the fault.

The logging rate may be selected. The digital output may be interrogated every *n* seconds and each depth logged, or the mean of *n* soundings may be logged at preset intervals. The analogue trace is not discarded. The surveyor will undoubtedly wish to monitor the bottom trends during the operation, and the trace is utilised as a check on the correct digitising of the bottom echoes. Various ways of indicating correct digitising are used. The most common consist of either a blocking pulse immediately following receipt of the leading edge of the echo (e.g. Kelvin Hughes) or the keying of a short pulse at a discrete interval after receipt of the echo (Atlas). The first produces the effect of a 'white line' between the bottom profile and the reverberation 'tails' (see Plate 6). The second is recorded as a 'ghost' line exactly following the seabed profile, about 20 mm below it. A 'normal' record with no white line, or a break in the fine 'ghost' line, indicates digitising failure.

An alternative method of digitising is provided by a pen-follower such as the d-Mac. The echo-sounder trace is placed on the instrument table and the seabed profile traced by a manually operated pen or graticule. At 'turning points' on the profile a button on the graticule is pressed and the co-ordinates of the points relative to the grid of the table are recorded on computer-compatible paper or magnetic tape. This method implies that the processing operation takes place in a properly equipped shore office and, of course, the table co-ordinates must be converted to those of the map-grid in use for the correlation of fixes on the trace with those on the track chart.

The Automated System
The positioning and depth data having been reduced to digital format, they now must be interfaced together and related to time before the information may be output. An extremely accurate crystal clock is usually used to activate the interrogation of the position and depth sensors at the required intervals in addition to providing a time-scale for the recorded digital data.

At this point the data are at the stage where an X-Y plotter could be controlled by the computer via the interface to produce a track-plot (and, if required, a plot of raw depth data). All that is now required is for a tape of tidal data to be 'read' by the computer and the appropriate reductions applied to the depth data, and the system is capable of outputting a bathymetric plot. Further refinement of the program

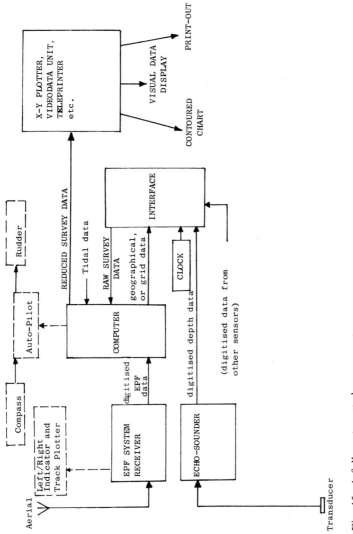

Fig. 45. A fully automated survey system

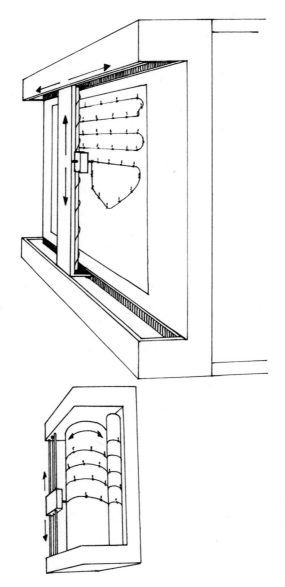

Fig. 46. The drum-plotter (left) and flat bed (right)

can cause contour-crossing points to be marked by characteristic symbols along the sounding lines, and the contours themselves may be plotted automatically following a least-squares matrix interpolation by the computer. Automation need not end here. By introducing required heading data and the required line-length and spacing into the program, and interfacing the compass and steering control to the system, a vessel may be steered automatically along predetermined lines.

A function diagram of a fully automated hydrographic survey system is shown in Fig. 45. (It should be noted that many sensors, not only the echo-sounder, may be interfaced with the positioning data.) Some common peripherals are also shown. Such a highly developed system is ideally suited for the use of radio-controlled sounding craft—a technique just beginning to earn serious consideration.

Computer Controlled Plotters
There are two types of X-Y plotter, the drum and the flat bed. Both incorporate a pen carriage. The drum plotter, like the track plotter supplied as a peripheral with all EPF systems, has the pen moving along the stationary carriage as one axis while the chart is driven by the drum around which it is rolled to provide movement along the orthogonal axis.

In the flat bed plotter, the chart is stationary, the pen movement is in one axis, the carriage movement in the other.

The drum plotter occupies less space than the flat bed but only the portion of the chart lying under the pen carriage can be inspected. In the case of the flat bed plotter, the whole chart area is open for inspection (see Fig. 46).

In conclusion, it may be noted that the precisely plotted and consistent style of the automatic plotter is usually acceptable to the client and the production of a separate fair sheet in addition to the original bathymetric sheet is unnecessary.

CHAPTER SEVEN

Relevant Formulae and Computations

7.1 Introduction

The hydrographic surveyor finds over a period that he is asked to acquire and process many different kinds of data as outlined in the opening chapter. Common to all is position, and while guidelines on this aspect can be (and are) usefully included here, the reader's familiarity with geodetic and grid computations is assumed.

In many cases, for example the determination of reduced depth or tidal stream rates and directions, simple arithmetic applied to the observed data is sufficient and no detailed explanation is required. In others, such as the assessment of sediment transport and studies for ecological purposes, reduction and correction tables are used together with empirical formulae and graphical plots to deal with the several interrelated parameters. The inclusion of instructions for these procedures without a great deal more detail in the text would be pointless, and the reader is necessarily directed to the Bibliography.

In this chapter, therefore, some brief notes are given on computations concerning position-fixing at sea, in amplification of parts of the relevant chapters. It must be remembered that many facets remain unexplained, such as the derivation of propagation velocities and calibration procedures for EPF systems. The reader is cautioned not to embark on field operations without the benefit of experienced supervision and the appropriate system handbooks.

7.2 The Construction of Lattices of Constant Subtended Angle for Position-fixing by Sextant

The principles of construction of arcs of constant subtended angle are illustrated in Fig. 7, p. 24 and in Fig. 47 below.

Given the co-ordinates of the shore stations (A and B) calculate:

(i) the length of the baseline AB;
(ii) the co-ordinates of the centre point of the baseline (C);
(iii) the co-ordinates of a second point (D) to enable the baseline perpendicular bisector to be plotted.

The centres of all the arcs required lie on this perpendicular bisector. The minimum and maximum subtended angles are estimated from a rough plot or existing chart of the survey area. Then calculate:

(iv) the lengths of the radii (R) of the selected arcs;
(v) the distance from C of the circle centres along the perpendicular bisector;
(vi) the co-ordinates of the circle centres.

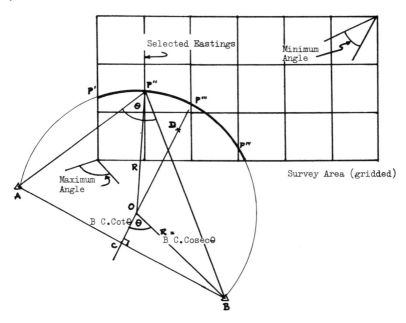

Fig. 47

If the survey area and shore stations are contained within the limits of the plotting sheet the arcs are then scribed using beam compasses.

If the plotting sheet is too small to accommodate the shore stations, the points at which the arcs intersect selected eastings (or northings, depending on which gives the greater angle of cut) must be computed, and the arcs scribed along curves placed over the intersections. Referring to Fig. 47: for an angle θ subtended along an arc P', P'', P''', by the shore control points A and B, values of eastings may be taken from the rough plot. The northings ordinate at which the arc cuts the chosen eastings is then calculated from the expression:

$$\Delta N = \sqrt{R^2 - \Delta E^2}$$

where ΔN is the difference of northings between the circle centre and P, and ΔE is the difference of eastings between the circle centre and the selected easting.

7.3   Determination of the Uncertainty of a Fix by Electromagnetic Position-fixing System

Every EPF system has a limit to the reliability of the read-out, imposed by instrumental imperfections and fluctuations in signal strength. This may be one or two hundredths or thousandths of a lane depending on the system and its operating frequency, range, etc.

Referring to Fig. 48, the root mean square error (E) in metres for a 65 per cent probability is given for a hyperbolic chain by the expression:

$$E = \cos \beta \sqrt{\sigma^2 W_1^2 \cos^2 \frac{\gamma_1}{2} + \sigma^2 W_2^2 \cos^2 \frac{\gamma_2}{2}}$$

where β is the angle of cut of the position lines (hyperbolae);
σ_1 and σ_2 are the standard deviations of the patterns I and II in metres obtained from the expression $\sigma_1 = \sigma W_1 \cos \gamma_1/2$ (and similarly for σ_2), where σ is the standard deviation in hundredths (thousandths) of a lane as described above;
W_1 and W_2 are the lane widths in metres of the two patterns on their respective baselines, i.e.

$$W = \frac{\text{velocity of propagation}}{2 \times \text{pattern frequency}};$$

γ_1 and γ_2 are the angles subtended at the ship's position by the Pattern I and II baselines.

The expression $W \cos \gamma/2$ is termed the Lane Expansion Factor and gives the width of the hyperbolic lane at the point of observation. In a two-range chain there is no lane expansion, and E in this case is given by the expression:

$$E = \sigma W \cos \beta$$

For a 95 per cent probability, the radius of the circle of uncertainty is $2E$.

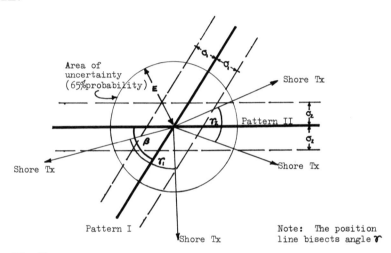

Fig. 48

7.4 The Conversion of Hyperbolic EPF System Lane Read-out to Geographical Co-ordinates

For the computation of latitude and longitude from system lane-readings the following data are required:

$\left.\begin{array}{cc} \phi_I & \lambda_I \\ \phi_{II} & \lambda_{II} \\ \phi_{III} & \lambda_{III} \end{array}\right\}$ The latitude and longitude of the shore transmitters I, II and III

$N_1 \; N_2$ The hyperbolic lane readings of Patterns 1 and 2 at the ship's position P

Let $(R + x)$, R and $(R + y)$ be the ranges of the ship at P from the shore stations I, II and III respectively.

Now, if shore stations I and III are enslaved to the master station II,

$$N_1 = \frac{\text{I–II} + R - (R + x)}{\lambda_1} = \frac{\text{I–II} - x}{\lambda_1}$$

and
$$N_2 = \frac{\text{II–III} + R - (R + y)}{\lambda_2} = \frac{\text{II–III} - y}{\lambda_2}$$

where λ_1 and λ_2 are the wavelengths of the Pattern 1 and 2 signals. These are known.

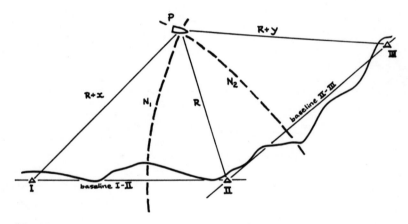

Fig. 49

If the stations are independent:

$$N_1 = \frac{R - (R + x)}{\lambda_1} = \frac{-x}{\lambda_1} \quad \text{and} \quad N_2 = \frac{R - (R + y)}{\lambda_2} = \frac{-y}{\lambda_2}$$

Thus,

$x = \text{I–II} - \lambda_1 N_1$ (enslaved system)

$\qquad\qquad\qquad\qquad$ or $\quad -\lambda_1 N_1$ (independent system)

and

$y = \text{II–III} - \lambda_2 N_2$ (enslaved system)

$\qquad\qquad\qquad\qquad$ or $\quad -\lambda_2 N_2$ (independent system)

Let $\phi_{P'}$, $\lambda_{P'}$ be the latitude and longitude of a trial point P', taken from a rough plot, or estimated from the previous fix.

$$\text{Let } d\phi = \phi_P - \phi_{P'} \quad \text{and} \quad d\lambda = \lambda_P - \lambda_{P'}$$

Knowing the spheroid in use, the semi-minor axis (a) and eccentricity (e) are known.

Compute the spheroidal radii of curvature at P' ($\rho_{P'}$ and $v_{P'}$):

$$\rho_{P'} = \frac{a(1 - e^2)}{(1 - e^2 \sin^2 \phi_{P'})^{3/2}} \qquad v_{P'} = \frac{a}{(1 - e^2 \sin^2 \phi_{P'})^{1/2}}$$

Compute distances and azimuths from P' to I, II and III (i.e. R', $R' + x$ and $R' + y$, and α_I, α_{II} and α_{III}). Using any long-line formula, compute:

distances between P' and I, II and III (R'_I, R'_{II} and R'_{III}); azimuths from P' to I, II and III ($\alpha_{P'I}$, $\alpha_{P'II}$ and $\alpha_{P'III}$); and azimuths from I, II and III to P', ($\alpha_{IP'}$, $\alpha_{IIP'}$ and $\alpha_{IIIP'}$). The distances must be calculated to an accuracy equal to or better than the accuracy of the EPF system (say 1 m).

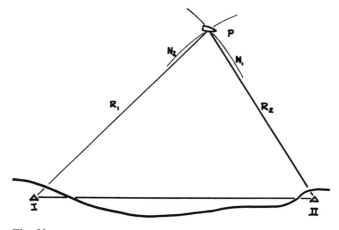

Fig. 50

The corresponding 'observed' distances are $R + x$, R and $R + y$. Let $R = R'_{II} + dR$: the 'observed' distances are then $R'_{II} + dR + x$, $R'_{II} + dR$ and $R'_{II} + dR + y$. Derive variation of co-ordinate equations:

$$-\rho_{P'} \cos \alpha_{P'I} \sin 1'' \, d\phi + v_I \cos \phi_I \sin \alpha_{IP'} \sin 1'' \, d\lambda$$
$$= R'_{II} + dR + x - R'_I$$

$$-\rho_{P'} \cos \alpha_{P'II} \sin 1'' \, d\phi + v_{II} \cos \phi_{II} \sin \alpha_{IIP'} \sin 1'' \, d\lambda$$
$$= R'_{II} + dR - R'_{II} = dR$$
$$-\rho_{P'} \cos \alpha_{P'III} \sin 1'' \, d\phi + v_{III} \cos \phi_{III} \sin \alpha_{IIIP'} \sin 1'' \, d\lambda$$
$$= R'_{II} + dR + y - R'_{III}$$

which become:

$$-\rho_{P'} \cos \alpha_{P'I} \sin 1'' \, d\phi + v_{I} \cos \phi_{I} \sin \alpha_{IP'} \sin 1'' \, d\lambda - dR$$
$$= R'_{II} + x - R'_{I}$$
$$-\rho_{P'} \cos \alpha_{P'II} \sin 1'' \, d\phi + v_{II} \cos \phi_{II} \sin \alpha_{IIP'} \sin 1'' \, d\lambda - dR$$
$$= 0$$
$$-\rho_{P'} \cos \alpha_{P'III} \sin 1'' \, d\phi + v_{III} \cos \phi_{III} \sin \alpha_{IIIP'} \sin 1'' \, d\lambda - dR$$
$$= R'_{II} + y - R'_{III}$$

Solve for $d\phi$, $d\lambda$ and dR.

Then, if $d\phi$, $d\lambda$ exceed the EPF system resolution, apply them to $\phi_{P'}$ and $\lambda_{P'}$ to obtain a new trial point and repeat. If $d\phi$ and $d\lambda$ are smaller than the system resolution,

$$\phi_P = \phi_{P'} + d\phi \quad \text{and} \quad \lambda_P = \lambda_{P'} + d\lambda$$

7.5 The Conversion of Two-range EPF System Lane Read-out to Geographical Co-ordinates

Symbols are as used in para. 7.4. In this case, the lane readings indicate the number of half wavelengths between the ship and shore stations and the ranges may be derived directly:

$$R_1 = \frac{N_1 \lambda_1}{2} \quad \text{and} \quad R_2 = \frac{N_2 \lambda_2}{2}$$

The triangle IPII can be solved, all three sides being known.

7.6 The Construction of Lattices for use with Two-range and Hyperbolic EPF Systems

A lattice for a system in the two-range mode will consist of a series of concentric (spheroidal) circles representing lines of zero phase at intervals of one half wavelength, centred on the shore transmitting stations. Therefore, given the velocity of propagation of the radio signals and their frequency, the wavelength and lane-width can be calculated. For plotting at the scale of the survey the mean scale factor and natural scale of the survey are applied in the computation of the plotting radii. (See Figs 10 and 11.)

A lattice for the hyperbolic mode is begun in exactly the same way, the hyperbolae being constructed by joining the intersection points of the two-range circles whose differences give a constant value, e.g. circles numbers 20 and 15, 21 and 16, 22 and 17, etc. It is convenient for range circles to meet on the baselines. Therefore, if the baselines are not an exact number of half wavelengths long, it is usual to increase the radius of the first lane-circle around one of the shore stations by an amount equal to the fraction remaining. For example, if the baseline is 1000·30 lanes in length, the first circle at one of the stations would have a radius equal to 1·30 lanes and the others would have radii of 2·30, 3·30, and so on.

If the survey area is too distant from the shore stations for the use of beam compasses, the circles are scribed using curves placed along co-ordinated control points as in the case of sextant-arc lattices. The numerous control points required justify the use of a computer, programmed using the formulae in paras 7.4 and 7.5.

Bibliography

Literature on specific aspects of hydrography is comparatively scarce. The available standard works have been mentioned wherever the text warrants throughout the book. Those titles are listed below together with additional references, with brief notes on their particular merit as further reading.

Definitive Works

The Admiralty Manual of Hydrographic Surveying. Hydrographic Department, Ministry of Defence, 1965 *et seq.*

In its two volumes, undoubtedly the standard work on hydrographic surveying. Vol. I deals almost entirely with positional control, including the calculation and preparation of plotting sheets and lattices. Vol. II has been published chapter by chapter since 1965 to deal with the individual subjects and operations of hydrography, such as tides and tidal streams, sounding, marks and marking, and sweeping.

Both volumes are based on the production of the nautical chart. They are traditional in character and naturally have a heavy service bias. Since 1965 a great many advances have been made, notably in the form of remote sensing (oceanographical probes, side-scan sonar, underwater beacons) and position-fixing by satellite and long range EPF, and the manual necessarily stops short of these. No attention is given to costing, of course, nor to the many industrial applications of hydrographic techniques.

Sea Surveying. Ingham: John Wiley, 1974
A treatise oriented to survey management covering all aspects of measurement and investigation at sea to a medium, somewhat academic level. Not intended as a manual, the book attempts to complement the *Admiralty Manual*'s coverage and report the advances in evidence since 1965.

Electromagnetic Distance Measurement. Burnside: Crosby Lockwood, 1971
Among the several texts on EDM, this work stands out for its treatment of marine EPF systems in the context of e.m. waves and instrumentation generally.

Radio Aids to Maritime Navigation and Hydrography. International Hydrographic Organisation Special Publication No. 39, IHB, Monaco, 1965
As a directory of systems, this book is concise but inevitably somewhat outdated. However, its supplements, together with the *IHB Review* and manufacturers' literature (see below) repair the breach.

The Use of Hi-Fix in Hydrography. Decca Survey Ltd
Though produced for company staff and users of the system, this work deserves mention for its authoritative treatment of all matters relating to Decca Hi-Fix (and similar systems) including the siting and calibration of chains and the calculation and plotting of lattices.

The International Hydrographic Review. Journal of the IHB, Monaco, 6-monthly
Through articles submitted by contributors and reprinted from other journals and conference proceedings, this journal constitutes an admirable source of up-to-date information on all instrumentation and techniques, e.g. Omega, Transit, sonar of all types, geophysical and oceanographical data acquisition and interpretation methods.

The Admiralty Manual of Tides. Hydrographic Department, Ministry of Defence, 1941
The standard work on tides. Somewhat heavy going for the average

mind, but definitive, authoritative and comprehensive. The theory of tides is given more simply in the *Admiralty Manual of Hydrographic Surveying* Vol. II and *Sea Surveying*.

The Admiralty Tidal Handbooks. Hydrographic Department, Ministry of Defence
> No. 1. *The Admiralty Semi-Graphic Method of Harmonic Tidal Analysis* (1959)
> No. 2. *Datums for Hydrographic Surveys* (1960)
> No. 3. *Harmonic Tidal Analysis for Short Period Observations* (1964)

These booklets are the accepted authority for all tidal work required of the hydrographic surveyor.

Fundamentals of Sonar. Horton: US Naval Institute, 1957
> Acoustic wave propagation and the principles of sonar instrumentation are fully detailed in this volume, in certain particulars unnecessarily so from the surveyor's point of view.

Underwater Investigation Using Sonar (1966) and *Sonar in Fisheries— a Forward Look* (1967). Tucker: Fishing News. *Applied Underwater Acoustics*. Tucker and Gazey: Pergamon, 1967
> Provide a more simple, relevant and readable explanation for the surveyor and are highly recommended.

Instruction Manual for Obtaining Oceanographical Data. US Naval Oceanographical Office Publication No. 607, 1968
> The acquisition and reduction of observations for salinity, temperature and all other commonly measured seawater parameters, together with seabed and seawater sampling, are clearly described.

Manufacturers' literature is frequently explicit on matters relating to instrumentation. Particularly recommended are the following:

The Plessey Co. Ltd. Paper by A. G. Marshall: 'Development of the MRB 201/301'.

The Magnavox Company (UK agents S. G. Brown Ltd, Watford). Theory of satellite position-fixing and all aspects of the Transit method.

Sercel, Divn. Radio Location, Paris. Principles of the Toran EPF Systems.

Simrad AS, Oslo. Echo-sounder systems for hydrography and fisheries research.

Background Reading

Oceanology Today. Barton: Aldus Books, 1970
  A readable account of the scope of industrial activity offshore by the editor of *Offshore Services*.

Physical Geography of the Oceans. Cotter: Hollis & Carter, 1965
  A short but comprehensive treatment of the terminology and nature of the marine environment.

Encyclopaedia of Oceanography. Fairbridge, Ed: Reinhold, 1966
  Complete, concise and praiseworthy.

Oceanography—Readings from Scientific American. W. H. Freeman, 1971
  Readable and beautifully illustrated. Excellent general coverage of all branches of oceanography.

Developments

Of the journals commonly subscribed to in the UK, perhaps the best selection would include the *IHB Review, Offshore Services, Dock and Harbour Authority* and the *Hydrographic Journal.*

  Conference proceedings are a fruitful source of news on modern developments and trends. The following proceedings are particularly recommended for the papers mentioned:

NORSPEC 72 (North Sea Spectrum)—The proceedings of a conference on the ships, materials, equipment and the problems involved in the exploitation of the North Sea. Troup, Ed: Thomas Reed, 1971
  'Hydrographic surveying in the North Sea' (Glen)
  'Feasibility studies for dredging projects' (Sargent)

'Decca Survey high precision positioning facilities in the North
Sea' (Parkes)
'How deep is the sea?' (Cloet)
'Some practical applications and limitations of high definition
depth scanning sonars' (Cook)
'Acoustic techniques for geological studies' (Gauss)

Dredging. Institution of Civil Engineers, 1968
'Towards automation in inshore hydrographic surveying' (Hales
and Cloet)
'The use of tracers to determine infill rates in projected dredged
channels' (Crickmore)

First Marine Geodesy Symposium. US Government Printing Office,
1967
'Apollo ship navigation accuracies and requirements for marine
geodesy' (Byrd)
'Underwater acoustic positioning, principles and problems'
(Tyrrell); 'Applications' (Spiers)
'Application of hyperbolic radio systems to marine geodesy' (Dean)
'Acoustic navigation: surface and subsurface' (Cline)
'Absolute position determination of sonar beacons' (Engel, Lev
and Gelb)
'Precise positioning with a laser theodolite' (Cunningham)
plus many interesting papers on geodesy and geophysics.

XIIIth International Congress of Surveyors, Commission 4. Fédération
Internationale des Géomètres, 1971
All papers are of direct relevance to hydrographic surveying. Par-
ticularly interesting papers are:
'Hydrographic and tidal information for deep draught ships in a
tidal estuary' (White)
'Hydrographic automation—a progress report (Roberts and Weeks)

**DIAL
READ OUT
UNIT**
**USED IN
REMOTE MODE**

**DIGITAL RANGE
INTEGRATOR**
USED IN MASTER MODE

1. The Tellurometer MRB 201. (Courtesy The Plessey Company Limited)

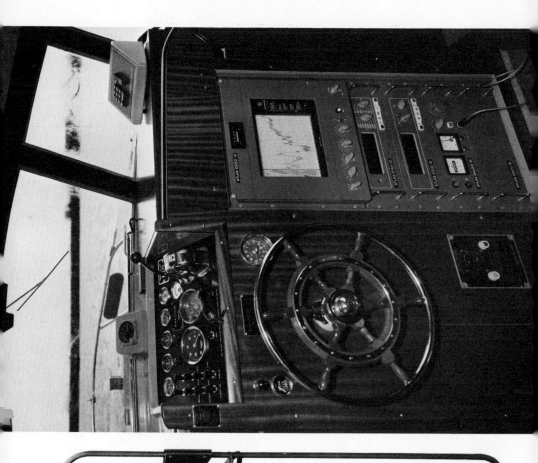

11. A Savonius rotor type of recording current meter. The graphic recorder is shown at top (protective hemisphere removed), above the current speed rotor and direction vane. (Courtesy Hydro Products Division of the Dillingham Corporation)

12. The Atlas Survey System SUSY 10 installed in a survey launch: (top to bottom) DAMA data manual input, DESO echosounder, DIRA digital range read-out, EDIG digital depth read-out, RALOG distance measurement unit, DACU data control unit. The computer and plotter are not shown. (Courtesy Fried. Krupp GmbH Atlas Elektronik)

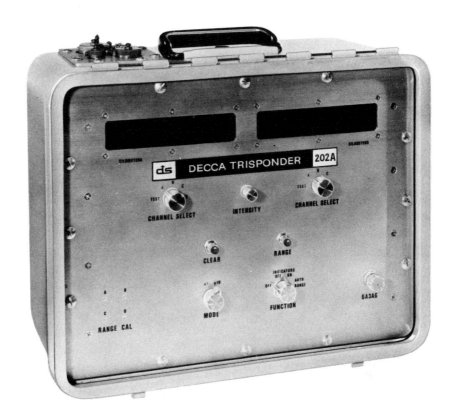

2. The Decca Trisponder Survey System Model 202A: (above) the DMU and (below) a 'remote' transponder. (Courtesy Decca Survey Limited)

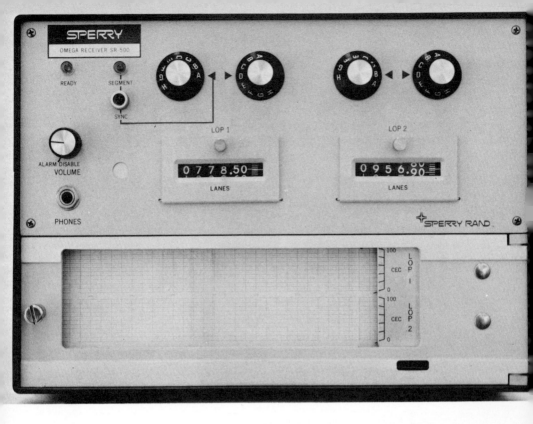

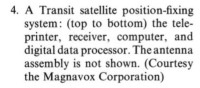

3. An Omega receiver. (Courtesy The Sperry Gyroscope Division, Sperry Rand)

4. A Transit satellite position-fixing system: (top to bottom) the teleprinter, receiver, computer, and digital data processor. The antenna assembly is not shown. (Courtesy the Magnavox Corporation)

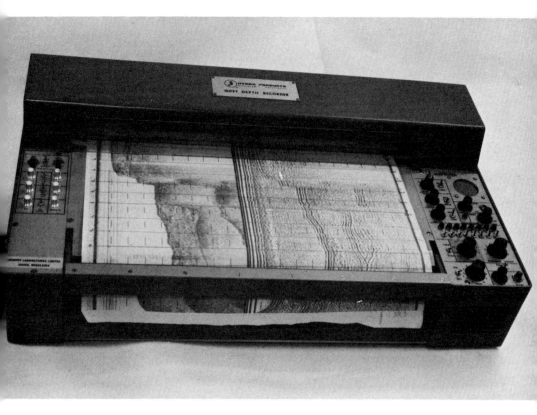

5. A Gifft precision depth recorder. The trace shows simultaneous records from seismic sources of different resolution and penetration. (Courtesy Wimpey Laboratories Limited)

6. (Left) Part of the trace from a Kelvin Hughes MS26A echo-sounder, showing the principal features and including a 'false' echo from a fish shoal.
(Right) Part of the trace from a Kelvin Hughes MS36 echo-sounder with digital output. The digitisation of the bottom echo is indicated by the 'white line'. Non-digitisation is indicated by the solid returns shown at extreme left.

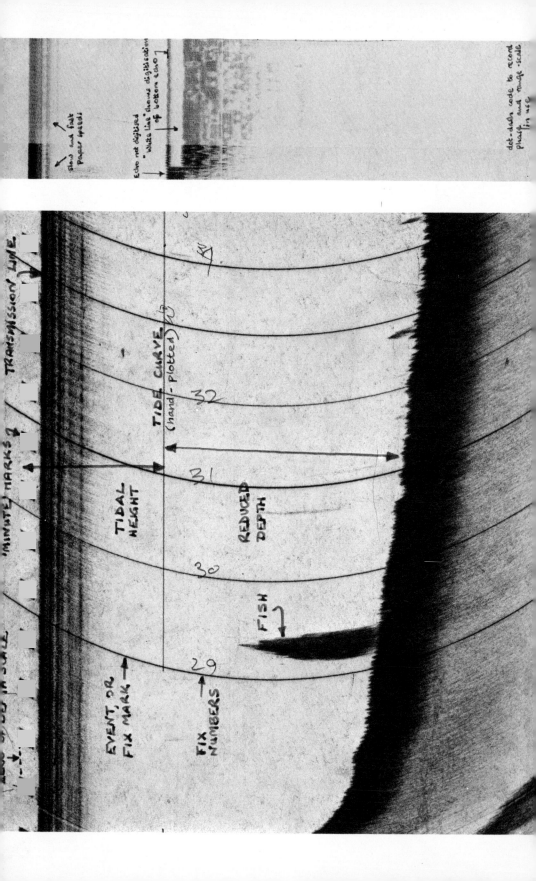

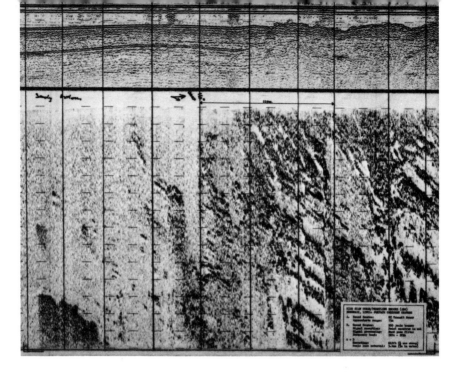

7. Simultaneous records from a high resolution seismic source and a side-scan sonar, showing rock outcrops. (Courtesy Wimpey Laboratories Limited)

8. A dual channel side-scan sonar record showing submarine cables traversing a rock outcrop. Parts of the cable are buried in areas of silty sand. (Courtesy Hunting Surveys Limited)

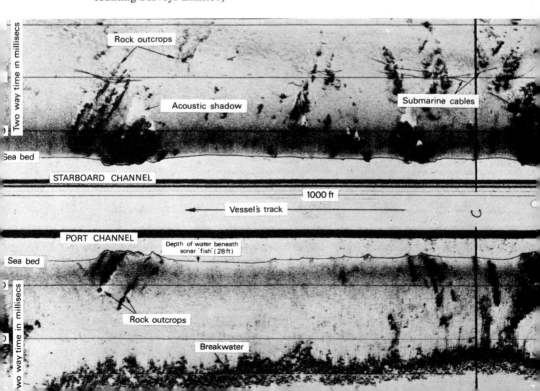

9. The Kelvin Hughes MS43 single channel side-scan sonar recorder. (Courtesy Kelvin Hughes Division of Smith's Industries Limited)

10. An underwater acoustic transponder beacon. (Courtesy Edo Western Corporation)

Index